TECHNICAL REPORT

T0302551

Workforce Planning and Development Processes

A Practical Guide

Georges Vernez, Albert A. Robbert,
Hugh G. Massey, Kevin Driscoll

Prepared for the United States Air Force

RAND PROJECT AIR FORCE

The research described in this report was sponsored by the United States Air Force under Contract F49642-01-C-0003 and FA7014-06-C-0001. Further information may be obtained from the Strategic Planning Division, Directorate of Plans, Hq USAF.

Library of Congress Cataloging-in-Publication Data

Workforce planning and development processes : a practical guide / Georges Vernez ... [et al].
p. cm.
Includes bibliographical references.
ISBN 978-0-8330-4006-0 (pbk. : alk. paper)
1. United States. Air Force—Personnel management. 2. Manpower—United States. I. Vernez, Georges.

UG773.W67 2007
358.4'1610973—dc22

2007000929

The RAND Corporation is a nonprofit research organization providing objective analysis and effective solutions that address the challenges facing the public and private sectors around the world. RAND's publications do not necessarily reflect the opinions of its research clients and sponsors.

Published 2007 by the RAND Corporation
1776 Main Street, P.O. Box 2138, Santa Monica, CA 90407-2138
1200 South Hayes Street, Arlington, VA 22202-5050
4570 Fifth Avenue, Suite 600, Pittsburgh, PA 15213-2665
RAND URL: http://www.rand.org/
To order RAND documents or to obtain additional information, contact
Distribution Services: Telephone: (310) 451-7002;
Fax: (310) 451-6915; Email: order@rand.org

Preface

The Air Force Materiel Command (AFMC) is establishing a capability to engage in comprehensive, analytically based management of its workforce to improve its ability to meet the ongoing and anticipated Air Force demands for products and warfighting capabilities.

As part of this effort, the AFMC Director of Personnel (AFMC/A1) asked the RAND Corporation to assist in identifying, analyzing, and preparing effective approaches for workforce planning and development. This report describes the multiple steps involved in analytically grounded force management processes and outlines the decisions that need to be made at each step, describes the types of processes and tools that others have used to support each step, and recommends how responsibilities ought to be divided between the various AFMC organizations.

This report should be of interest to all personnel managers and to functional and operational leaders throughout AFMC and its centers and laboratory directorates. It should also be of interest to anyone else interested in workforce planning and development.

The research reported here was sponsored by the AFMC Headquarters Directorate of Personnel (AFMC/A1), and was conducted within the Manpower, Personnel, and Training Program of RAND Project AIR FORCE.

RAND Project AIR FORCE

RAND Project AIR FORCE (PAF), a division of the RAND Corporation, is the U.S. Air Force's federally funded research and development center for studies and analyses. PAF provides the Air Force with independent analyses of policy alternatives affecting the development, employment, combat readiness, and support of current and future aerospace forces. Research is conducted in four programs: Aerospace Force Development; Manpower, Personnel, and Training; Resource Management; and Strategy and Doctrine.

Additional information about PAF is available on our Web site at http://www.rand.org/paf.

Contents

Figures

Tables

Boxes

Summary

With more than 100,000 military, civilian, and contractor personnel, AFMC is responsible for developing, acquiring, and maintaining most Air Force weapons and other military systems. Unlike the rest of the Air Force, most AFMC personnel are civilians (56 percent) or contractors (26 percent), many of whom have specialized technical skills. As technology changes, as new weapons are developed, and as battlefields' operating requirements change, so do AFMC's personnel requirements.

To address its current and future workforce needs, AFMC is working to strengthen its capabilities to engage in proactive, analytically based workforce planning and development. This practical guide to workforce planning and development is a contribution to this effort. It describes the four main steps involved in analytically grounded workforce planning and development, outlining the policy decisions that need to be made at each step and substep; describes the methods, data, and tools that others have used and/or need to develop to support each step; and recommends how responsibilities ought to be divided among Headquarters (HQ) AFMC, business units, and functional managers:

1. **Determine workforce demand (see pp. 7–23).** Determine what the workforce should be now and in the future in terms of its
   - size—the total number of positions needed
   - composition—the proportions of military, civilian, and contractor personnel
   - job competencies requied—the backgrounds each position or group of positions requires.
2. **Describe workforce supply (see pp. 25–30).** Describe the workforce as it is now and as it is projected to be in the future, given existing personnel policies and practices. This task requires describing the current and projected force in the same way as for the ideal current and future force demanded in step 1.
3. **Compare the demand with the supply (see pp. 31–36).** Identify gaps between what the workforce is and what it needs to be, and assess the options for filling any such gaps.
4. **Implement solutions (see pp. 37–39).** Implement changes to fill the gaps identified between workforce demand and supply, and administer them on an ongoing basis. This involves developing the processes, practices, and monitoring tools needed to assure that the force is managed on a day-to-day basis and over time to meet workforce requirements.

AFMC does not need to spread its planning and development efforts over its entire workforce. Rather, we recommend that AFMC focus on its core business units: the three product centers, three logistics centers, nine laboratory directorates, and three test units that account for 70 percent of its workforce. Within these units, AFMC should focus only on the positions that are central to the mission's respective units. Such positions may include nonleadership positions. (See pp. 9, 16–17.)

Consistent with AFMC's concept of operations, we generally recommend assigning the responsibilities for workforce planning and development to the business units. HQ AFMC's primary role should be to guide the process and provide support and assistance to the business units and decide on trade-offs that may need to be made between business units. Similarly, functional managers have an advisory role. (See pp. 3–5.)

Before it can fully engage in the workforce planning and development outlined in this document, AFMC needs to strengthen its analytical capabilities, both at HQ AFMC and in its business units. (See p. 5.)

Acknowledgments

We are grateful to many people who helped guide and support this research. In particular, we thank Leif Peterson (AFMC/A1), who identified the need for the study. Maggie Grace and Lt Col Cheryl Malone provided valuable advice and guidance throughout. Those in AFMC/A1 who provided valuable advice or responded to our many questions included

Carolyn Bowen
Capt Sara Cooper
John Day
Col Ruth Ellis
Lt Col Paula Goode
Gena King
Gary Kuzienski
Lt Col Steve Niemantserdriet
Marti Paul
Maj David Penson
Pat Shama
Col Dave Taylor

Others in AFMC met with us to discuss their force management activities, including

Joe Berube (AFMC/A3)
Col Paul Copp (ASC/EN)
Robert DiTommaso (AFMC/EN)
Robert Hill (AFMC/PK)
Lisa Mazur (AFMC/A2/5)
Capt Lauren Miller (AFMC/A4)
Rosalie Benackin Roberts (ASC/HR)
Sharon King-Roberts (AFMC/FM)
Polly Sweet (AFRL/DP)

We are also grateful to Lionel Galway and Craig Moore, our RAND colleagues, who shared their knowledge and experience with various aspects of force management, and Michael Thirtle and Ellen Pint, who provided very thoughtful reviews of this document.

Finally, we appreciate the editorial assistance of Shelly Wiseman and the administrative assistance of Louis Ramirez.

Abbreviations

AAC	Air Armament Center
AAC/CC	Commander, Air Armament Center
AF/A1M	Air Force Directorate of Manpower and Organization
AFMC	Air Force Materiel Command
AFMC/A1	Director of Personnel, Air Force Materiel Command
AFMC/A2/5	Director of Intelligence and Requirements, Air Force Materiel Command
AFMC/A8	Director of Strategic Plans and Programs, Air Force Materiel Command
AFMC/CC	Commander, Air Force Materiel Command
AFPC	Air Force Personnel Center
AFPC/DPK	Directorate of Civilian Career Management, Air Force Personnel Center
AFRL	Air Force Research Laboratory
AFSC	Air Force specialty code
AFSLMO	Air Force Senior Leader Management Office
AFSPC	Air Force Space Command
C2ISR	command, control, information, surveillance and reconnaissance
CC	commander
DO	Directorate of Operations
DP	Directorate of Personnel
DT	development team
EN	engineering
ESC	Electronic Systems Center
FM	financial management

HQ	headquarters
LG	logistics
MEO	most efficient organization
NAF	numbered air force
OMB	Office of Management and Budget
PAF	Project AIR FORCE
PEO	program executive officer
PK	program management
PME	professional military education
PWS	performance work statement
SACOM	Sustainment and Acquisition Composite Model
SAF/AQX	Deputy Assistant Secretary of the Air Force for Management Policy and Program Management
SPO	system program office

CHAPTER ONE

Introduction

The Air Force Materiel Command (AFMC) is responsible for developing, testing, acquiring, delivering, and logistically supporting most Air Force weapons and other military systems. One of the largest organizations within the U.S. Air Force, it has more than 100,000 military, civilian, and contract personnel in multiple locations in the United States and around the world. Nearly half the civilians who work for the Air Force work for AFMC, and of these, most choose to spend most of their careers with AFMC. Moreover, AFMC personnel have specialized skills somewhat unique to the organization because its component laboratories and test, acquisition, and logistics centers have technical missions that are unique within the Air Force.

AFMC's personnel needs are expected to change as technology changes and as new weapon systems are developed and operational requirements change. Nevertheless, past efforts to identify future force requirements for the acquisition or logistics labor force have typically justified the status quo.[1] Staff we interviewed at various AFMC components reported that they typically filled vacancies with staff having skills similar to those of the previous jobholder. The organization recognizes the need to approach workforce planning and development more strategically. Indeed, some AFMC units have already increased their involvement in force management, expanding their human resources functions to include analytical and planning functions, using indicators to monitor the characteristics (composition, occupations, education, experience, and the like) of their workforces, and seeking to identify broadening and training requirements. However, these efforts are neither comprehensive nor coordinated across units, and the overall AFMC organization has limited capability to evaluate its workforce as a whole or to guide its development.

AFMC is establishing a capability for comprehensively and analytically managing its workforces. Toward this end, RAND Project AIR FORCE (PAF) was asked to assist in identifying, analyzing, and preparing effective approaches for workforce planning and development, including addressing such questions as how the process should be organized and what the size and qualifications of staff should be to support deliberate force management.

As part of this project, this report describes the multiple steps (see Figure 1.1) involved in analytically grounded force management processes. It outlines the decisions that need to be made at each step, describes the types of processes and tools that others have used and/or need

1 For instance, see U.S. Air Force (2002).

to be developed to support each step, and suggests how responsibilities ought to be divided between the various AFMC organizations.

Approach

In preparing this document, we reviewed the literature on workforce planning and development, including the considerable body of work PAF has already done (Crawford, 2001; Sullivan, 2002; Emmerichs, Marcum, and Robbert, 2004a, 2004b; Gates, Ebner, and Keating, 2006; Vernez et al., 2004; Robbert et al., 2005). The more comprehensive workforce planning and development approach outlined in this document combines many of what we judged to be the best practices offered by this previous work.

We also conducted interviews with staff at Headquarters (HQ) AFMC (including at the Directorate of Operations; the Directorate of Personnel (A1); and the functional managers for logistics, financial management, program management, and engineering and sciences (formerly LG, PK, FM, and EN); Aeronautical Systems Center; and Air Armament Center (AAC) to identify current workforce planning capacity, practices, and issues. These interviews were conducted in winter 2005.

Overview of Workforce Planning and Development

Workforce planning and development, succinctly referred to as *force management*, is about getting the right number of people with the right set of skills and competencies in the right job at the right time. Figure 1.1 illustrates the four major steps involved in workforce planning and development and the range of inputs, managerial and analytical, needed to carry them out:

1. **Determine workforce demand.** Describe what the workforce should be now and in the future in terms of its
   - size—the total number of positions needed
   - composition—the proportions of military, civilian, and contractor personnel
   - job competencies required—the backgrounds required by each position or group of positions.
2. **Determine workforce supply.** Describe the workforce as it is now and as it is projected to be in the future, given existing personnel policies and practices. This task requires describing the current and projected force in the same way as for the desired current and future force in step 1.
3. **Compare the demand with the supply.** Identify potential gaps between the current workforce and the desired workforce, and assess the options for filling any such gaps. Management has three primary levers for filling gaps:
   - new accessions into the workforce—bringing new personnel into the appropriate area within AFMC, either new hires or lateral moves from another governmental organization

- changing the compositional mix of military, civilian, and contractor personnel—getting the skills and expertise needed by selecting the appropriate staff regardless of whether they are military personnel, civilians, or contractor employees
- force development through successive job assignments and education and training—developing the skills and expertise needed by working with existing AFMC staff and providing the education, training, and experience needed.

4. **Implement solutions to meet demands.** Fill the gaps identified between demand and supply. This involves developing the processes, practices, and monitoring tools needed to assure that the force is managed on a day-to-day basis and over time to meet workforce requirements.

The force management processes and programs put in place in step 4 ought to be monitored. This involves developing appropriate indicators to ensure that implementation is proceeding as planned and evaluating whether the programs put in place are having the effects desired.

As shown in Figure 1.1, implementing each of the four workforce planning and development steps requires the active involvement of the line managers to consider trade-offs and make administrative and policy decisions. It also requires considering national security and Air Force long-term strategic directions and plans. Finally, it requires the collection of information, the use of personnel data files, and the use or development of analytical tools including projection and simulation models.

The remainder of this document elaborates on the tasks involved in each step, identifying the following:

- policy or administrative decisions to be made
- methods, data, and tools
- assignment of responsibilities between the human resource managers at AFMC's headquarters, centers, laboratory directorates, and test units and its functional managers at its headquarters.

Also, and for the purpose of this document, the term *job competencies* is used to include the set of education, training, skills, experiences, and personal traits and expertise that are relevant to perform a specific job effectively. It includes what the Air Force has referred to at different times as *occupational* or *technical* competencies, as well as enduring, institutional, and leadership competencies. At times, the term can indicate a broad, integrated set of these elements. At other times, individual elements, such as specific academic degrees or work experiences, may be referred to collectively as *competencies*.

Assignment of Responsibilities

Consistent with the 2004 concept of operations, *Roles and Responsibilities of AFMC Center Line Direct Reports and Functionals* (AFMC, 2004), we generally propose that the responsibili-

Figure 1.1
Workforce Planning and Development Process

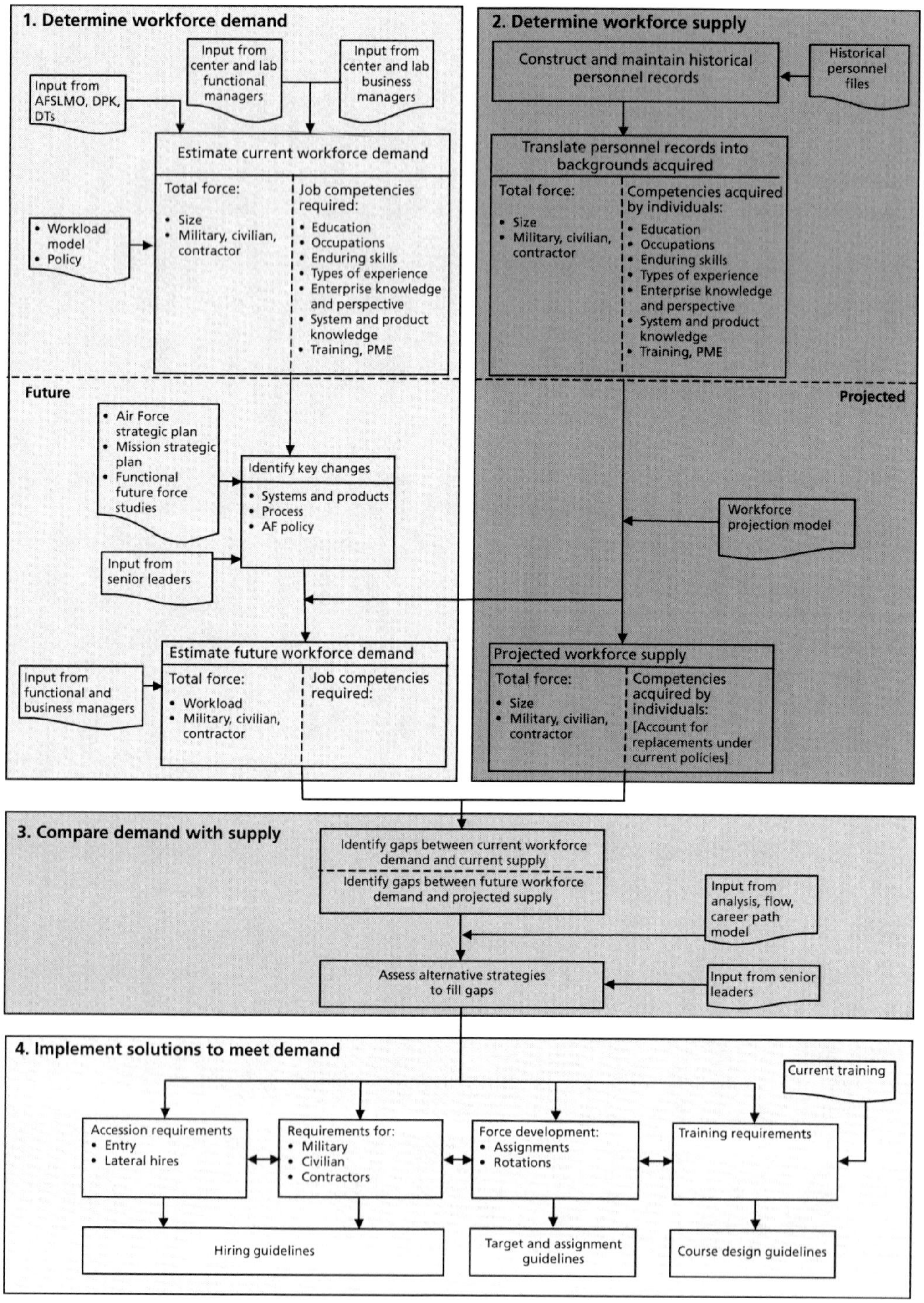

RAND *TR408-1.1*

ties for force management be assigned to AFMC line organizations or business units, such as centers and laboratories. As a result of the recent return of program executive officer (PEO) authority to AFMC, the concept of operations gives product and logistics centers' commanders the "authority over program execution and the resources (including labor resources) required to deliver and sustain war-winning, air capable weapon systems" (AFMC, 2004). In labor force management, this means that line commanders and/or directors' responsibilities include assessing workload requirements, identifying position and skill requirements, making reassignments within line organizations, and making the decision to fill a position from outside or via lateral transfer or promotion. Functional managers at HQ AFMC and within the center staffs have an advisory role regarding these matters but also retain control over reassignments across line organizations.

In this context, the primary roles of HQ AFMC would be to guide the process, support and assist the business units, and function as the moderator of workload and resource tradeoffs that may need to be made between business units.

In this document, reference to HQ AFMC generally refers to either AFMC/A1, AFMC/A8M, or both. Although these organizations are currently separate and have different reporting lines, they are complementary, and we recommend merging them into a single staff activity eventually. Such a merger would help integrate the manpower, development, and personnel functions, making the task of managing the workforce easier.

Our interviews with staff at both HQ AFMC and some centers indicated that few, if any, of the center human resource management organizations and AFMC/A1 presently possess the analytical capabilities to engage in deliberate force management as this document describes. AFMC expects to develop these capabilities over time, beginning with the acquisition of an analytical capability at AFMC/A1.

Organization of the Report

This report is organized around the first four of the five force management steps described above. Chapter Two addresses the question of defining both existing and future workforce demands. Chapter Three lays out the tasks needed to describe the current and projected workforce (the supply) in the same terms as the demand. Chapter Four then describes the range of analyses needed for identifying gaps between supply and demand, assessing the competencies of the workforce, and testing alternative strategies for addressing workforce shortcomings. Chapter Five summarizes the types of processes and guidance that need to be developed to guide daily hiring decisions and the development and training assignments of both military and civilian personnel. We conclude with some overarching observations and recommendations. The appendixes present our survey instruments.

CHAPTER TWO

Determining the Demand

Two main tasks for determining the workforce demand are (1) estimating *current* workforce requirements and (2) estimating *future* workforce requirements (Figure 2.1). Each of these is further divided into the following subtasks:

- size—total number of positions needed
- composition—shares of military, civilian, and contractor personnel
- job competencies required—competencies that specific positions or groups of positions require.

Estimating Size, Composition, and Competencies Required for the Current Workforce

Although interrelated, the three subtasks required for estimating the size, composition, and job competencies for the current workforce are driven by different considerations and require different analytical approaches and tools.

Estimating Total AFMC Workforce Size

The primary outcome of this subtask should be the number of total staff (military, civilian, and contractor personnel) each line organization and AFMC as a whole require to effectively and efficiently meet the workload demands the Air Force makes on the command. Aggregate manpower constraints are often set at Air Force, Department of Defense, and even congressional levels, thereby limiting manpower flexibility. AFMC's independent determination, however, serves two main purposes: First, it allows AFMC to make a convincing business case for the optimal size of each of its units and its total force relative to aggregate demand for inclusion in budget requests. Second, it informs trade-off decisions between AFMC business units when manpower allocations are limited, as is typical. To keep the estimate up to date, this process should be performed annually.

Figure 2.1
Determine Workforce Demand

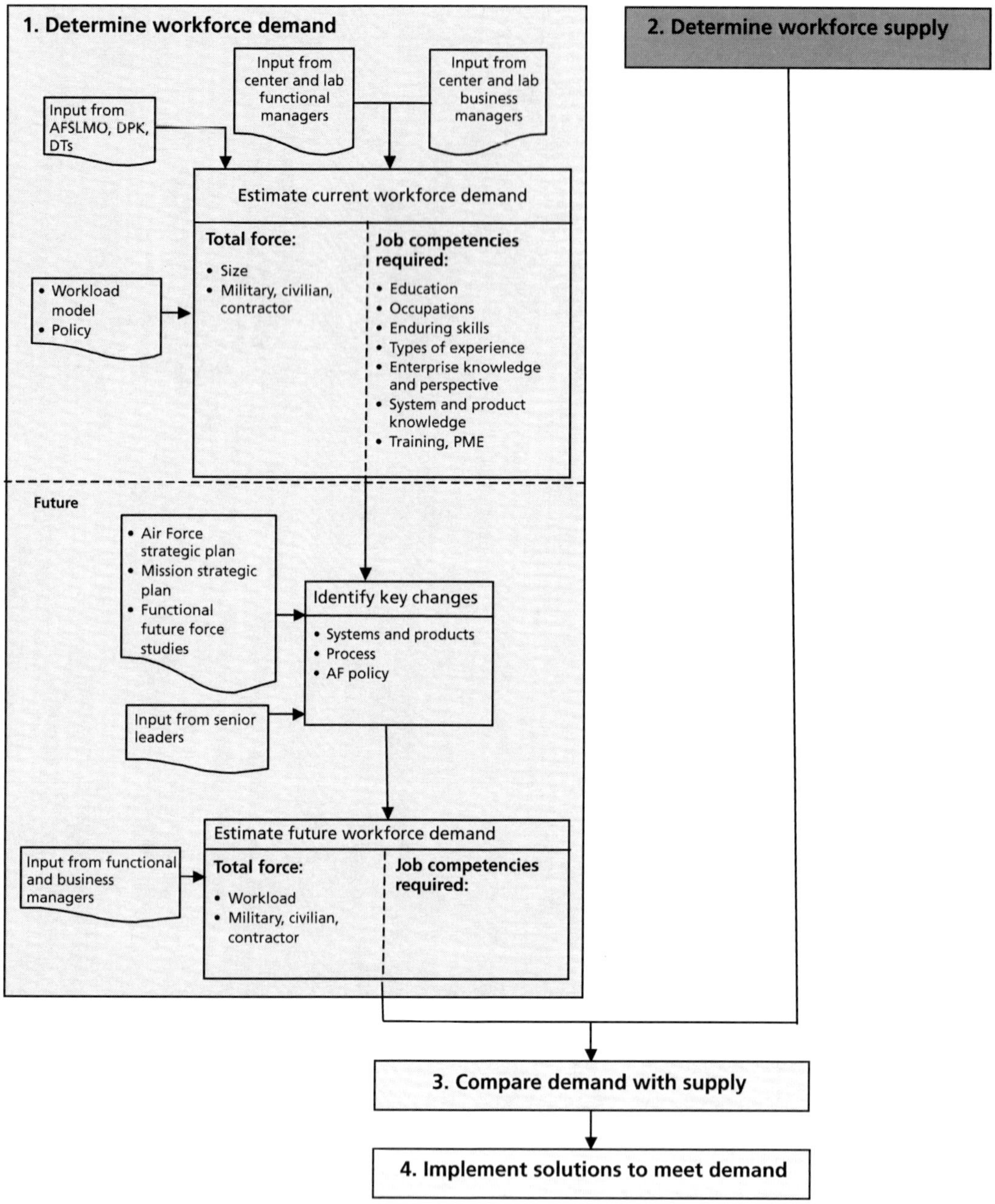

RAND *TR408-2.1*

Policy Decisions Needed. AFMC needs to determine the organizational elements for which it will strategically manage its workforce. It is an extremely large organization that develops different products and fulfills a variety of functions. But as noted in the summary, the core of its activities is carried out in a set of quasi-independent business units: ten laboratories, three test centers or units, three product centers, and three logistics centers (see Table 2.1). Each of these business units deals with different products at different stages of development, acquisition, and/or sustainment. Product ideas and prototypes are initially developed at the laboratories or by firms under contract to the Air Force. Once the decision is made to procure a new product, its acquisition is moved to a product center that is typically organized around system program offices (SPOs), which may be dedicated to a single major weapon system or a grouping of similar minor products. The workload for each product goes through a life cycle that begins at a low level, peaks, and eventually declines as the product becomes obsolete, so workforce requirements vary over time. In turn, the logistics centers maintain inventories of spare parts and repair (or contract to repair) the purchased products and modify (often modernize) them over their respective life cycles. Test centers are involved in ongoing testing, as needed, for each product.

We recommend that AFMC workforce planning and development focus on these core business units, and all subsequent references to business units in this document are to them. Together, these units include 70 percent of AFMC's organic (military and civilian) workforce and 70 percent of its contractor workforce. Force management of other AFMC activities, including the base support wings, can be effectively done within the context of Air Force–wide processes and programs.

Methods, Data, and Tools. Modeling is needed to estimate the workforce size required to handle the workload. Various techniques have been developed to make such estimates. They all implicitly or explicitly (via a mathematical relationship) associate some measure(s) of workload with the total number of workers required. Ideally, the measure(s) of workload should be related to the outcomes of the business units, such as the number of airplanes to be acquired, the number of bombs to be produced, or the number of engines to be upgraded. But measures of workload cannot always be used; such may be the case for research and development in the laboratory directorates. In the latter case, educated judgment and experience in project development may have to substitute for a more formal analytical approach.

One alternative approach to estimating the required force size is the federal government's most efficient organization approach (Box 2.1).

Currently, AFMC uses the Sustainment and Acquisition Composite Model (SACOM) to estimate the required size of its workforce in product and logistic centers.[1] As currently designed, it has several weaknesses (Box 2.2). PAF, in conjunction with the Air Force Directorate of Manpower and Organization (AF/A1M) and the Air Force Manpower Agency, is assessing SACOM and will recommend changes to it or the use of another methodology.

Assignment of Responsibilities. Force size ought to be estimated for each core business unit separately and then be aggregated AFMC-wide. The necessary analytical work may either be centralized at HQ AFMC or be done at the business units, with the aggregation at HQ

[1] AFMC developed this model over several iterations and is in the process of refining it.

Table 2.1
AFMC Core Business Units

AFMC Business Units	Organic Workforce[a]	Contractor Workforce[b]	Air Force Base	Main Products or Activities
Acquisition Centers	9,856	4,369		
Aeronautical Systems	4,579	1,351	Wright-Patterson, Brooks	Manned and unmanned aircraft systems; medical systems
Air Armament	990	591	Eglin	Air-to-air and air-to-ground missiles and munitions
Electronic Systems	4,287	2,427	Hanscom, Gunter, Lackland, Wright-Patterson	Air and ground command, control, information, surveillance and reconnaissance (C2ISR) systems, networks, information systems
Laboratory Directorates	5,345	1,591		
Air Vehicles	315	204	Wright-Patterson	Aircraft technologies
Directed Energy	534	77	Kirtland	Lasers, microwave technologies
Human Effectiveness	590	89	Wright-Patterson, Brooks	Human interface technologies
Information	810	169	Rome NY	Information handling technologies
Materials and manufacturing	485	96	Wright-Patterson	Materials technologies
Munitions	341	159	Eglin	Conventional munitions technologies
Propulsion	485	308	Wright-Patterson, Edwards	Turbine and rocket engine technologies
Sensors	707	133	Wright-Patterson, Hanscom, Rome	Sensor technologies
Space Vehicles	591	169	Kirtland, Hanscom	Military satellite technologies
Office of Scientific Research	155	55	Arlington, Va.	Basic research
HQ AFRL	332	132	Wright-Patterson	Laboratory headquarters

Table 2.1—Continued

AFMC Business Units	Organic Workforce[a]	Contractor Workforce[b]	Air Force Base	Main Products or Activities
Logistics Centers	32,856	4,884		
Ogden	9,871	1,454	Hill	F-16, A-10, C-130, intercontinental ballistic missiles, space C2ISR systems, landing gears
Oklahoma City	12,247	1,240	Tinker	KC-135, E-3, bombers, turbine engines
Warner-Robins	10,738	2,190	Robins	C-5, C-141, C-130, F-15, helicopters, avionics systems
Test and Evaluation	5,854	5,197		
Arnold Engineering Development Center	278	2,186	Arnold	Wind tunnel and simulation testing
Air Force Flight Test Center	3,664	1,159	Edwards	Aircraft flight testing
AAC (46th Test Wing)	1,912	1,852	Eglin, Holloman	Munitions testing
Other AFMC	22,754	6,952	Multiple	HQ AFMC, field operating agencies, air base wings, specialized centers
Total AFMC	76,665	22,993		

SOURCE: Air Force Manpower Data, September 2004.

[a] Including military and civilian personnel.

[b] Including on-site contractor force associated with the respective business units.

Box 2.1
The Most Efficient Organization Approach (OMB Circular A-76)

The most efficient organization approach, which is used for competitive sourcing studies, specifies the procedures that the federal government must follow when it considers outsourcing a function that is currently being provided by civil service employees. The procedure requires developing a performance work statement that specifies what needs to be accomplished without articulating how the work should be performed. Then, a detailed work plan must be prepared for accomplishing that work. The information on personnel requirements may include the grade level and the job title or series. The justification for the manpower requirements requires comparing the estimated characteristics of the workforce with those of the current workforce. Any major differences are to be discussed in the documentation.

Gates, Ebner, and Keating (2006) assessed the potential of this approach for demand analysis and concluded that the approach is consistent with effective workforce planning and could be used to determine workforce size requirements, including those for functions performed mainly by civilian personnel.

This approach would be most useful for essentially similar functions that are performed at multiple locations or in multiple organizations. The development of the most efficient organization approach at one location could serve as a benchmark for manpower requirements in similar activities at other locations.

Box 2.2
The Sustainment and Acquisition Composite Model (SACOM)

SACOM is AFMC's current tool for quantifying the aggregate manpower needed in acquisition and logistics centers. In 2004, the model estimated that AFMC required 15 percent more personnel than it had to perform the then-current workload.

SACOM describes the work to be performed and correlates—via regressions—a range of workload factor scores to the number of people required. The workload factors include such items as (1) the quantity and level of required reporting (2) the volatility of user requirements (3) the magnitude of contracting activity (4) the amount of interaction with other government agencies, and (5) the amount of management and technical oversight required.

Despite being an improvement over its predecessor, the Product Support Resource Model (PSRM), SACOM has several weaknesses: a focus on processes rather than on outputs, a relationship between factor scores and manpower requirements that ignores how well the observed organizations actually perform, a lack of understanding of what drives the workload requirements and the needs for them, and inadequate recognition that manpower needs change over a weapon acquisition and sustainment life cycle.

AFMC. Experience suggests that it is best to develop manpower estimates at the business unit level, where familiarity with the mission can be tempered with an appreciation of resource constraints. Under this option, the assignment of responsibilities would be as follows:

- **HQ AFMC**
  - provides directives and guidelines to business units to prepare their respective size estimates
  - provides access to data needed
  - provides technical assistance for application of estimating models
  - reviews units' estimates and recommends trade-offs across units as needed
  - advocates for inclusion of force size requirement in the program objective memorandum
- **centers and laboratory directorates**
  - estimate the current required workforce
  - provide rationale for estimates
  - negotiate trade-offs with HQ AFMC
  - advocate for manpower requirements.

Estimating Workforce Composition: Military, Civilian, and Contractor Personnel

The next task is to estimate the composition of the force in terms of the shares of military, civilian, and contractor personnel. These compositional requirements should be reviewed periodically, as experience dictates and/or policy changes.

The composition of the workforce currently varies significantly among AFMC's core business units (Table 2.2). For instance, among product centers, the Aeronautical Systems Center has the highest proportion of organic (military and civilian) staff, with only 23 percent on-site contractors. Both AAC and the Electronic Systems Center (ESC) rely more extensively on contractors (37 and 36 percent, respectively).

Similarly, the share of contractors ranges from a low of 13 percent in the Directed Energy and Human Effectiveness Laboratory directorates to 39 percent in the Propulsion Directorate, and it ranges from 25 percent to 89 percent in the test and evaluation centers/units.

The share of contractor personnel varies not only across centers but also across occupational areas. For instance, the share of contractors among engineers and scientists ranges from 8 percent in the Aeronautical Systems Center to 55 percent in AAC and to 78 percent in ESC.

Variations in the use of contractors across centers and within the same occupations across centers are so large that it is hard to see whether they reflect an optimal composition that is attributable to differences in labor requirements or local labor market conditions or whether they simply resulted from historical conditions that may no longer prevail.

When line managers in the AAC were asked what the optimum share of contractors would be, their answers ranged broadly, from a low of 20 to a high of 80 percent. The low figure reflected respondents' concerns for continuity and institutional memory. The high figure reflected the view that contractors could do most of the functions of an SPO with the exception of the director and key financial, security, and contracting positions.

Similarly, the proportions of military and civilian personnel vary across AFMC business units. At one extreme, the Arnold Engineering Development Center and the Oklahoma City Logistics Center are only 3 percent military, while at the other, the Air Force Flight Center is 38 percent military. On the other hand, 50 percent or more of the personnel in AFMC business units are civilians. Exceptions include ESC and the 46th Test Wing, which are less than 50 percent civilian, and the Arnold Engineering Development Center, which is nearly all civilian.

Policy Decisions Needed. Well-defined criteria need to be developed to determine which positions ought to be military, which civilian, which contractor, and which any of these. Positions that *must* be filled by military personnel may include those that require combat and/or field experience, provide valuable system know-how for transfer into the field, or require command of military personnel.

Table 2.2
Military, Civilian, and Contractor Distribution by AFMC Component

AFMC Component	Total Authorized (no.)	Military (%)	Civilians (%)	Contractors (%)
Acquisition centers	14,225	26	43	31
Aeronautical Systems	5,930	27	50	23
Air Armament	1,581	14	48	37
Electronic Systems	6,714	28	35	36
Laboratory directorates	6,936	17	60	23
Air Vehicles	519	9	51	39
Directed Energy	611	30	57	13
Human Effectiveness	679	33	54	13
Information	979	11	72	17
Materials and Manufacturing	581	9	74	17
Munitions	500	17	52	32
Propulsion	793	8	53	39
Sensors	840	13	71	16
Space Vehicles	760	17	61	22
Office of Scientific Research	210	14	60	26
HQ AFRL	464	28	44	28
Logistics centers	37,740	4	83	13
Ogden	11,325	5	82	13
Oklahoma City	13,487	3	88	9
Warner-Robins	12,928	5	78	17
Test and evaluation	11,051	27	26	47
Arnold Engineering Development Center	2,464	3	8	89
Air Force Flight Test Center	823	38	37	25
AAC-46th Test Wing	3,764	28	23	49
Total AMFC	99,658	21	56	23

SOURCE: Air Force Manpower Data, September 2004.

Positions that should not be filled by contractors include those involving inherently governmental activities, such as oversight, finance, and contracting responsibilities, as well as positions that require interservice or governmental interactions or that might create a conflict of interest for contractors. Some of these positions may be either military or civilian.

Positions that *should* be filled by contractors involve unique technical requirements or temporary positions.

Although well-defined criteria, such as the ones above, can assist in determining a preferred compositional distribution of the workforce, it is likely that a majority of positions could just as readily be filled by either military, civilian, or contractor personnel. In the end, hence, determining the most appropriate workforce composition involves making judgments and trade-offs among the following:

- the need for maintaining flexibility[2]
- Air Force military and civil service manpower allocations
- trade-offs within program funding[3]
- full costs of the personnel
- market conditions.[4]

Methods, Data, and Tools. Several tools need to be developed to guide determination of the appropriate composition of the force:

- a *survey instrument* or other similar data-collection method to gather information on which positions require a military, civilian, contractor, or either[5]
- the expected availability of military personnel at various grades in various occupations
- a workforce flow model, calibrated for selected civil service occupations, to ensure that sufficient entry-level and junior positions exist to develop workers to fill more-senior positions
- labor market surveys in occupations that may be in short supply nationally or locally.

Assignment of Responsibilities. Tasks could be allotted as follows:

- **HQ AFMC**
  - develops criteria for determining whether positions should be filled by military, civilian, or contractor personnel
  - develops an instrument for surveying line and functional managers
  - obtains or develops labor market surveys
  - organizes by center commander and functional manager reviews
  - analyzes data and develops flow models

[2] For instance, the minimum share of contractors should in part be dictated by the need to maintain the flexibility to bring in experts rapidly, as fluctuations in the workload and the types of activities may require.

[3] Contractors are paid out of program funds.

[4] These may make it difficult to hire some technically skilled workers at civil service pay scales and/or at certain sites.

[5] These data should be gathered from the appropriate line managers and career field managers for each position, who should be asked to respond using well-defined, uniform criteria. Line managers and functional managers should review results of the survey.

- provides results to centers and laboratory directorates
- reviews and approves the workforce compositions of centers and laboratories
- **centers and laboratory directorates**
  - provide inputs on criteria for determining the compositional needs of positions
  - administer surveys
  - review survey results
  - determine workforce composition
- **functional managers**
  - provide inputs on criteria for determining the compositional needs of positions
  - review survey results.

Estimating Competencies for Current Jobs

Information on the competencies each position requires—such as previous occupations, academic education, training, and experience—should be updated periodically, as new systems are acquired or as new capabilities are developed.[6]

Policy Decisions Needed. The Air Force is currently in the midst of an evolving force development program that is designed to "develop strong occupational and enduring competencies" (U.S. Air Force, 2004). To date, the Air Force has collected information on competencies required for general officer and senior executive service positions and on occupational (AFSC) requirements for O-6 and plans to do so for selected civilian GS-15 positions. Presently, plans for collecting data on needed job competencies are limited to leadership military and civilian positions at the O-6/GS-15 and potentially the O-5/GS-14 levels. This leaves many positions, particularly technical ones, at the tactical and the operational levels for which information on required competencies will not be collected.

At the same time, some Air Force organizations and functional managers have been forging ahead by collecting their own information on the competencies their positions need. The Air Force Space Command (AFSPC) is the first to have completed its assessment of required job competencies and to have set force development requirements. Unlike the Air Force's corporate effort, AFSPC focused on the tactical level.

To determine required competencies for its positions, AFMC may choose to rely exclusively on the corporate Air Force effort described above. Or it may choose to complement this by covering the positions that will not be part of the corporate program. Assuming that AFMC chooses to do so, it will be necessary to decide which positions AFMC and its centers and laboratories should focus on and what competencies should be considered for these positions.

Deciding Which Positions to Focus on. AFMC's organic force is large, with more than 75,000 positions. But not all positions are equally important to AFMC's mission and its force planning and development. Hence, we recommend that AFMC limit itself to determining the competencies needed for military and civilian positions that are core to its mission. As noted earlier, research and development, testing, acquisition, and logistics are core to the AFMC

---

[6] Although it is preferable to estimate and validate the competencies current jobs require before estimating those for future jobs, this task can be bypassed if the vision of the competencies that jobs will require in the near future differs significantly from those of current jobs.

mission, and the occupations within these areas are generally more technical than most others in the Air Force. As shown in Table 2.3, the majority of positions in these occupations are in AFMC, so these are the types of positions on which it should focus its force management efforts.

For such positions as those in Table 2.3, AFMC should gather information about needed competencies for all grades and for all military, civilian, and contractor positions. While AFMC should avoid duplicating the efforts of the Air Force–wide force development program, the command should ensure that the information the program collects includes material of interest to AFMC (see below).

Choosing Competencies to Focus on. Most jobs have multiple educational, training, occupational, work, and battlefield experience, as well as other requirements. It is not manageable to deal with all these requirements equally just as it is not optimal to focus on only one or two of them. Although it is possible to assess each position without reference to a predetermined list of competencies, it would be desirable to develop such a list to ensure consistency and to keep the number of dimensions manageable.

Table 2.3
Occupations in Which Air Force Civilians Fill 50 Percent or More of the Positions in AFMC (Officer-Equivalent Specialties)

AFSC Code	Description	Air Force Civilians in AFMC (%)	Air Force Officers in AFMC (%)
43t	Biomedical laboratory	88	15
62e	Development engineer	83	41
43e	Bioenvironmental engineer	78	26
21a	Aircraft maintenance	77	11
36m	Mission support	76	10
21r	Logistic readiness	75	9
63a	Acquisition management	69	49
61s	Scientist	65	25
64p	Contracting	60	30
43b	Biomedical scientist	58	19
65w	Cost analysis	54	39
65f	Financial management	54	18
60c	Program director	50	60

SOURCE: Air Force Manpower Data, September 2004.

The competencies of potential interest to AFMC fall into the following categories:

- academic education (academic area, degree)
- professional military education (PME)
- training (specific courses provided by the Air Force, AFMC, the centers, and/or the laboratories)
- occupational background
- technical specialization within occupation
- prior job experience
- product systems
- functions (e.g., requirements, plans and programs, war planning)
- organizations (e.g., joint command, Headquarters Air Force, major command, center, laboratory)
- command (squadron, group, wing, or equivalent)
- AFSC prefix
- grade
- enterprise knowledge (e.g., joint, Air Force, AFMC, specific center or laboratory)
- leadership and management competencies
- language(s)
- performance level.

Whether a position should be filled by military, civilian, or contractor personnel could be added to this list instead of being treated as a separate (albeit important) dimension for the purpose of data collection.

The relative importance of the above competencies may differ across the centers and laboratory directorates, so the list should be developed in consultation with line and functional managers.

Methods, Data, and Tools. Determining the competencies needed for the selected positions requires surveying line and functional managers. The survey instrument for this should contain the key background dimensions of interest and should ask respondents to rate the relative importance of each competency or the proficiency level for each required skill.

The survey may be administered to line supervisors and career field managers, either online, by mail, or in focus groups. RAND Corporation experience suggests that the last may be preferable because it allows respondents to discuss the competencies that may be required and permits consensus (Vernez et al., 2006). Regardless of the method used to administer the survey, however, senior leaders, including the AFMC, laboratory, and center commanders, should review its results.

Box 2.3 illustrates the types of competency information sought for three different types of positions RAND has previously studied. Appendix A includes the worksheets we used for gathering information on two of these types of positions. Each emphasized different types of competencies.

Assignment of Responsibilities. The required competencies can be defined centrally or at the center and laboratory directorate level via a survey of all relevant positions. Since there is likely to be overlap across centers and laboratory directorates, a flexible, centralized approach may be preferable that would include all center- and laboratory-specific competencies, such as experience on specific systems or products. Tasks could be allotted as follows:

- **HQ AFMC**
  - decides on the position and job competencies of interest
  - designs survey instrument(s)
  - analyzes survey results
  - organizes AFMC senior leader reviews of survey results
  - distributes survey results to centers and laboratory directorates
  - builds and maintains database of job background requirements and makes the data accessible to centers and laboratory directorates
- **centers and laboratory directorates**
  - provide inputs for selection of positions and competencies of interest
  - provide inputs on design of survey instrument(s)
  - administer survey
  - facilitate line managers' completion of the survey
  - review survey results
- **functional managers**
  - provide inputs for selection of positions and competencies of interest
  - provide inputs on design of survey.

Box 2.3
Backgrounds Included in Selected Previous Surveys in the Air Force

Senior Leaders	Rated Officers	13S Officers
1. Academic (level, field) 2. Position requirement - Command - Length of stay - Partnered position 3. Skill requirements (72 options, including all occupations, operational skills, functions) 4. Means of acquiring operational experience for mobility, information, and aerospace power employment 5. Future requirements 6. Enduring competencies (11 options) 7. Courses required 8. Language requirements 9. Additional information to prepare future candidates 10. Priorities regarding above	1. Operational experience (18 options) 2. Previously held AFSC prefix (17 options) 3. Prior functional experience (28 options) 4. Organizational experience (44 options) 5. Prior leadership experience (24 options) 6. Combinations, other experience 7. Grade 8. Academic degree (2 options) 9. PME and technical training (8 options)	1. Operational experience (11 options) 2. Previously held AFS prefix (12 options) 3. Prior functional experience (17 options) 4. PME, unique to position (open-ended) 5. Degree area, unique to position functions (undergraduate and graduate) 6. Command experience (12 options) 7. Organizational experience (19 options) 8. Combinations, other experience 9. Grade

SOURCES: The survey sheets used for rated officer and for space and missile officer (AFSC 13S) positions appear in Appendix A. For an example of how the information collected can be used, see Vernez et al. (2006).

Estimating Size, Composition, and Competencies Required for the Future Workforce

Typically, a thorough understanding of the current composition of and competencies in the AFMC workforce would be the frame of reference from which to anticipate the probable size, composition, and competencies for the future workforce. As the 2006 Quadrennial Defense Review (QDR) outlines, the new defense strategy relies on transformation for its success. The reforms that this transformation entails emphasize jointness (e.g., the development of tailorable joint force modules), shift planning from a threat-based to a capabilities-based approach, and seeks to improve the use and acquisition of technology by taking full advantage of advances (Asch and Hosek, 2004). The last of these is integral to transformation, involving a revolution in the manner, speed, and effectiveness of industrial and commercial tasks and depends on the effects of advances in technology in computing, communicating, and networking (Asch and Hosek, 2004, p. 35). Rapid advances can be expected in sensors, communication, integration of information, surveillance and reconnaissance systems, precision-guided munitions, and directed energy weapons.

No one fully knows what these changes imply for workforce requirements. But, we can expect increasing emphasis on conserving active-duty positions for combat-essential activities, thereby increasing active-duty combat and combat support personnel and, hence, shifting combat support that is not combat essential to civilian or contractor personnel. It also may mean longer assignments for officer and enlisted personnel; while this would allow them to learn a job in greater depth it would also mean fewer opportunities to gain multiple types of experience. These and other potential implications of military transformation need to be systematically identified and analyzed for their consequences for force size, composition, and development.

Determining the demand for the future will require (1) identifying potential key changes in products, production processes, organization, or policy over a predetermined period and (2) estimating the resulting changes in the size, composition, and competencies required for the workforce.

Identifying Key Changes in the Future

Policy Decisions Needed. The first thing to determine is the period over which to anticipate workforce requirements. Five to six years may be too short, both because too little change may occur and because it is not enough time for significant alterations of the composition and skill mix of the workforce. Too long a period, say 20 to 25 years, may require predicting technological changes or the nature of threats and, hence, capability requirements, and such predictions have often been proved wrong in the past. A period of 10 to 15 years seems most appropriate. Most technological changes that are likely to occur within this time are already known, although their military applications may still be in development (as with directed energy) or may even not yet be in development.

Methods, Data, and Tools. Ideally, expected changes need to be identified in the following four areas, whether or not they imply a change in workforce requirements:

- technology and systems
- development, acquisition, or logistics processes
- organization
- Air Force and AFMC personnel policies.

Several techniques have been developed for dealing with the future, but nearly all these have three things in common: They all start with the known world or situation at hand; provide a disciplined, often stepwise, process that systematically engages experts' judgment; and consider uncertainty, that is, the probability that an expected event is actually going to take place (Winkler et al., 1998).

Techniques available for dealing with the future range from trend extrapolation to refocusing (outlining a vision of what might be that departs significantly from the present). AFSPC used the latter approach to define its requirements for space and missile officer (AFSC 13S) positions.[7] Perhaps the best-known technique for such projections is the Delphi method (Dewar, 2002). It relies on expert opinions, asking each expert to list potential changes he or she anticipates. An iterative process in which the experts provide feedback on each other's views eventually leads to consensus on key potential scenarios (visions) for the future and the uncertainty attached to them.

The process outlined below, which we tested in the context of AFMC Armament Enterprise, adapts aspects of these techniques by involving experts, but only after analysts have developed a set of potential changes:

1. Develop an initial list of potential changes, drawing from a careful review of Air Force and AFMC strategic plans, known technologies in development, and capability development requirements and complemented by consultations with Air Force staff involved in the capability development process. A key ingredient of this task and the next is to describe the expected changes as specifically as possible.
2. Discuss and review the list of potential changes with senior AFMC leaders (AFMC/CC, AFMC/A2/5, and laboratory and center commanders). They should add or delete items from the list, further specifying the nature of the potential change and assessing the probability that the change will occur, as well as the likely period within which it will occur. This review can occur in a one-day session involving all parties or over a series of individual meetings with each key senior leader. This process can be facilitated by the preparation of a *protocol*, such as the one in Appendix B. This group should also consider alternative future scenarios, i.e., sets of individual changes that are most likely to occur in concert.
3. A final list of potential changes is prepared based on the inputs from senior leaders. This list ought to be as detailed as possible. Box 2.4 shows such a final list that developed for the Armament Center.

---

[7] AFSPC chose not to survey its staff about requirements. Instead, it developed a vision of the competencies that 13S officers ought to acquire, focusing primarily on the mix and depth of systems and tactical experience.

Box 2.4
Elements of Change for Armament Enterprise

Product changes	Longer-range weapons
	Smaller, more-accurate weapons
	Inclusion of data links in weapons
	Weapons with constant communications with C2ISR
	Weapons that hit moving targets precisely
	Directed-energy weapons
Acquisition process changes	More integrated system-of-systems acquisitions
	More joint acquisition directed by Office of Secretary of Defense
	Increased interactions with operations
	Increased reliance on support contractors
Organizational changes	PEO restructuring
	Dedication of 10–15 percent of workforce to enterprise integration instead of program offices
	Grouping of SPOs

Assignment of Responsibilities. Same as assignment of responsibilities for translating key changes into the size, composition, and competencies required for the future workforce (see below).

Estimating the Future Workforce Demand

This task involves anticipating the workforce competencies that the expected changes in products; development, production, acquisition, or logistics processes; organization; and policies imply. Again, there is no alternative but to rely on the expert judgment of those who are close enough to the work that has to be done but high enough in the organization to have a strategic view. In our experience, this is found at the level of line and functional managers at the centers and laboratory directorates (Emmerichs, Marcum, and Robbert, 2004).

Policy Decisions Needed. None are needed.

Methods, Data, and Tools. The following activities can facilitate estimating workforce changes that are due to the identified changes in product technology, processes, or policies by the task described earlier:

- Show line and functional managers the distribution of current positions among the key competencies identified in the earlier job requirement task. For instance, if composition (military, civilian, contractor), occupational distribution, and/or academic background are the key dimensions of interest, the distribution of the force along them can be shown to respondents on a spreadsheet or, alternatively, as a bar chart.
- Ask respondents to indicate, below this current distribution of the workforce, how they expect it to change to address each of the previously identified product or process changes. A spreadsheet to facilitate this task is illustrated in Appendix A. In addition, respondents should be asked how each of the changes considered affects their workload.

This process should proceed from bottom up and should involve supervisors and directors at the highest level with sufficient workforce cognizance, such as the SPO level at product centers, the project level at the laboratory directorates, and the product groupings at the logistics centers. Laboratory and center commanders' and relevant functional managers' inputs should likewise be sought.

The results of this survey of line and functional managers need to be aggregated to develop a set of profiles of the future force requirements based on the change scenarios developed earlier. These future profiles can then be compared to the projected workforce profiles to identify gaps and, in turn, to guide personnel policies and programs, such as targets for hiring new personnel.

Similarly, expected changes in workload need to be translated into changes in the size of the workforce using the type of models described earlier in this chapter (Boxes 2.1 and 2.2).

Assignment of Responsibilities. Tasks could be allotted as follows:

- **HQ AFMC**
  - develops and oversees the process for estimating future workforce requirements
  - reviews and adjusts lists of potential key product, process, organizational, and policy changes developed by centers and laboratory directorates
  - sets up and moderates senior AFMC leaders' reviews of the lists of potential changes
  - reviews estimates of how potential product, process, and policy changes will affect workforce requirements that the centers and laboratory directorates have developed
  - aggregates the resulting estimates of future workforce requirements and develops a profile of the future AFMC-wide workforce
- **centers and laboratory directorates**
  - identify potential product, process, organizational, and policy changes over the next 10 to 15 years
  - ask line and functional managers to estimate changes in workforce distribution required along each key dimension and for each of the key changes
  - review estimates of how potential product, process, organizational, and policy changes will affect workforce requirements
- **line and functional managers**
  - estimate how potential product, process, organizational, and policy changes will affect workforce requirements.

CHAPTER THREE

Describing the Supply

The next step in the workforce planning and development process is to describe the workforce as it is now and as it is likely to be in the future, given existing personnel policies and practices, in the same dimensions used to describe the ideal current and future force. It involves three tasks: (1) constructing and maintaining historical personnel record files for both military and civil service civilians, (2) translating personnel records into the competencies staff members have acquired over their careers, and (3) projecting the current workforce over the same period selected for determining future workforce requirements (Figure 3.1).

Constructing and Maintaining Historical Personnel Records

To profile the supply side of the workforce, it is first necessary to build a historical file that contains the personnel records of the career of each AFMC military and civilian employee—not only within AFMC but also within the Air Force as a whole. This historical file should include all the information needed to translate individual experience into the language of workforce composition and competencies as defined in the previous step.

Policy Decisions Needed

The Air Force Personnel Center (AFPC) maintains and regularly updates personnel records for both military and civilians (Box 3.1).[1] Although extensive, these records may not contain all the information needed to determine all the relevant competencies AFMC and its centers and laboratory directorates deem important. For instance, our interviews with staff at the Armament Center and the Directed Energy Directorate suggested that more-detailed information may be needed on academic background and experience with specific products or systems than is currently being recorded. Similarly, participation in the training that AFMC's centers and laboratory directorates offer is not consistently recorded on personnel files. If more information needs to be collected on the competencies the workforce has acquired, AFMC will have to decide whether to do so, and if so, for what competencies. Collecting additional information on the competencies that military and civilian personnel have acquired and timely

[1] AFPC does not maintain records on contractors; hence, the competencies of the contractor workforce have to be collected or obtained from the federally funded research and development center or agencies through which contractors are hired.

recording of that information are resource intensive. Hence, AFMC and its centers and laboratory directorates may need to seek a balance between the desire to be as comprehensive as possible in identifying the job competencies required on the demand side and the costs of collecting the matching information on competencies actually acquired on the supply side.

Figure 3.1
Determine Workforce Supply

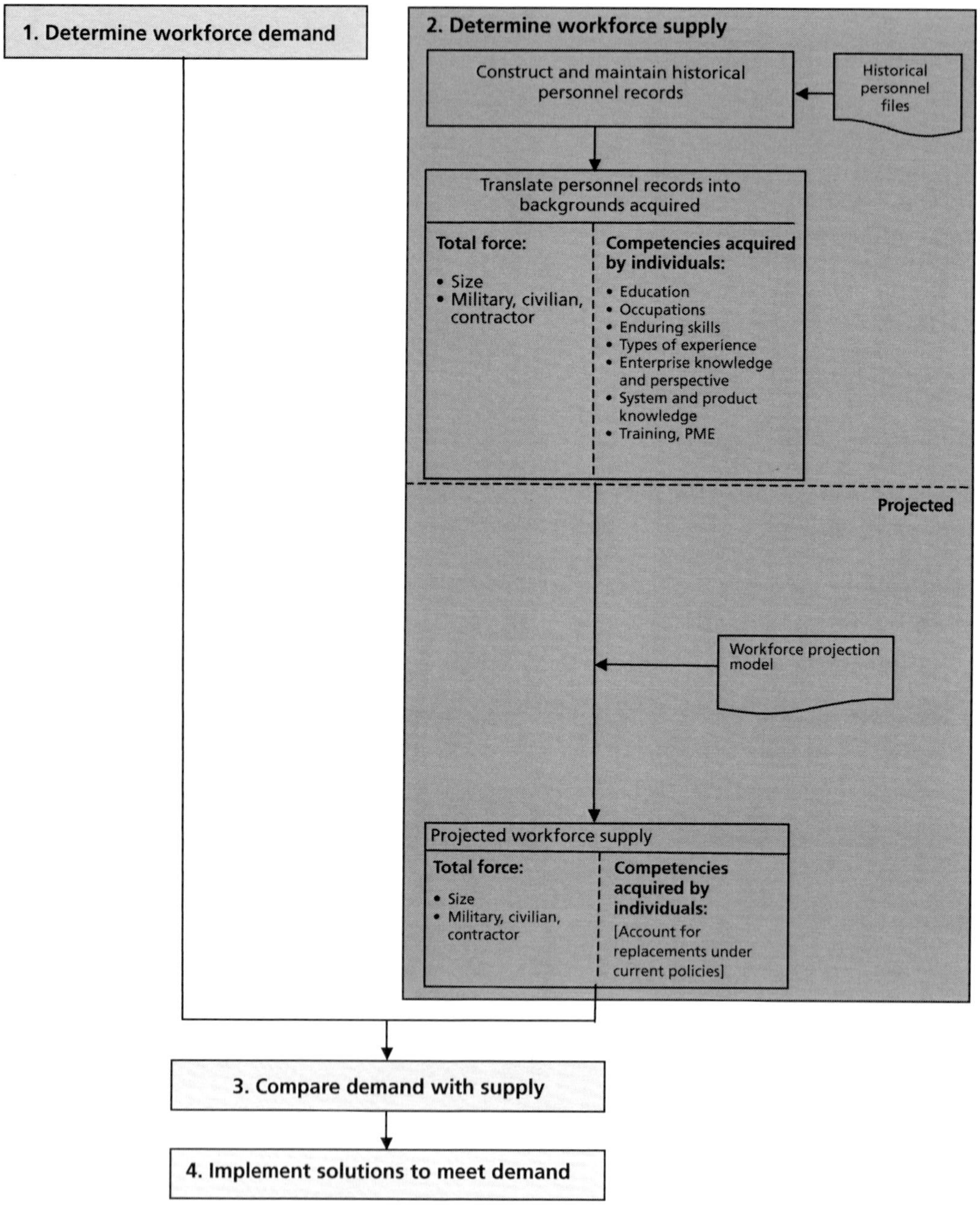

RAND *TR408-3.1*

Box 3.1
Content of Civilian Personnel Records

Civilian personnel records contain individual histories in the following areas relevant to workforce planning and development:

Personal and demographic information:	Reserve and/or veteran status, prior military service, date entered into civil service, retirement plan, and security clearance
Promotion history (up to 20):	Ratings pay plan, grade, awards, and dates
Educational history (up to 4):	Institution, degree or certificate, credit hours, instructional program, and dates
Languages (up to 4):	Proficiency in listening, speaking, reading, and writing
Training history (up to 4):	Type of training, course identification, method, hours of instruction, date completed, how funded, and cost
Job history (up to 25):	Career field, occupational series, and skill codes (16 data elements) Acquisition career position type and level (if applicable) Months of experience in each position Federal agency, major command Installation or location Employment type (civil service, military, nonfederal, etc.) Position level (in Air Force organization structure), organization functional code Supervisory status Work schedule (full time, part time, seasonal, etc.) Pay plan and grade Appointment type (permanent, career, temporary, etc.) Training program identification (if applicable)

Similar records are kept for military personnel, including education; PME; languages; and for each position ever filled by an individual, the AFSC, command level, organizational type and kind, unit, organizational structure name, location, and functional category.

Methods, Data, and Tools

HQ AFMC already has access to military and civilian historical data files and can use these to build extracts for the current AFMC workforce. As noted above, these extracts may require additional information on competencies of interest only to AFMC and that are not identifiable using the information currently recorded in personnel files.

Assigning Responsibilities. Tasks could be allotted as follows:

- **AFPC**
  - maintains and updates military and civilian historical personnel data files
- **HQ AFMC**
  - obtains and updates extracts of historical records for current AFMC military and civilian workforce
  - arranges to collect complementary information on competencies of interest, as defined by the competencies the jobs demand
  - enters complementary information on competencies into its personnel data files
  - provides centers and laboratory directorates access to AFMC personnel data files

- **centers and laboratory directorates**
  - make and maintain extracts of the historical records of their workforces
  - collect complementary information on competencies their workforces have acquired, to be entered into the AFMC and their personnel data files.

Translating Personnel Records into Competencies Acquired

Personnel records, as currently kept, do not necessarily map directly with the set of competencies used to determine demand. Some translation of the information in the language of competencies demanded is necessary.

Policy Decisions Needed

None are needed.

Methods, Data, and Tools

Translating personnel records into the competencies acquired involves developing heuristic computerized algorithms that cumulatively assign the appropriate acquired competencies to each individual throughout his or her career. Box 3.2 illustrates the approach RAND used to translate the job positions of space and missile (AFSC 13S) officers into a set of acquired competencies. The outcomes of this task will be new data files containing a complete set of the competencies that military and civilian personnel in AFMC have acquired. These files on the competencies of the current AFMC workforce should be updated annually to include any additional competencies that individuals may have acquired during the preceding year.

Assignment of Responsibilities

Tasks could be allotted as follows:

- **HQ AFMC**
  - develops and applies algorithms to translate current and aged personnel records into the competencies acquired; doing this centrally would assure consistent application of the same rules across all AFMC units
  - develops and updates historical files of competencies the workforce has acquired and makes them available to all appropriate AFMC units
  - provides analytical support to centers and laboratory directorates
- **centers and laboratory directorates**
  - provide inputs for developing the acquired-competencies algorithms
- **functional managers**
  - provide inputs for developing the acquired-competencies algorithms.

Box 3.2
Illustrative Algorithms Translating Space and Missile (AFSC 13S) Officer Personnel Records into Background Acquired

We used historical records (back to 1975) from the AFPC to identify the education, training, and experiences that space and missile officers had acquired. These records, however, could not be used as is. To credit officers with the appropriate tactical, operational, strategic, command, and organizational experiences they had acquired, these records had to be translated based on the various AFSC, unit, functional duty, and other codes available in the files. Because codes and organizations changed over time, care had to be taken to keep track of them.

For each job held, each officer was credited with the appropriate experience according to the following rules:

- Academic credit was awarded according to academic codes.
- PME credit was awarded according to training codes.
- Work experience fell into five main categories:
  1. Tactical experience (missile, satellite command and control, spacelift, surveillance, and/or warning) was awarded according to the appropriate operational AFSCs and/or assignments to operational space or missile units at the wing level or below, including Cheyenne Mountain Air Force Base operational centers (i.e., missile warning and space control centers).
  2. Specific experience was awarded according to prefixes received.
  3. Functional experience was awarded according to assigned functional duty (e.g., J-3, Directorate of Operations) in group-level and above organizations, functional AFSCs, functional codes, and major functional organizations (e.g., Air Intelligence Agency).
  4. Command experience was awarded according to C-prefix and group or wing command AFSCs (e.g., 10C, 91W).
  5. Organizational experience was awarded according to organizational and command codes.

The above general principles were written into a heuristic set of rules for computerized awarding of experience each officer acquired at each job held throughout his or her career. This task required extensive knowledge of the Air Force occupational classification system, as well as of its history.

SOURCE: Vernez et al. (2006).

Projecting the Current Workforce

So that the profile of the expected future workforce can eventually be compared to the competencies needed in the future, the current workforce needs to be projected *assuming that current policies would remain the same over the period specified in determining the demand.* What this means is that accession and hiring practices should be assumed to remain unchanged. And so they should be for retirement, lateral entry (for civilians), promotion, separation, and career development practices throughout the projected period.

Policy Decisions Needed

None are needed.

Methods, Data, and Tools

This task requires the development of a *workforce projection model.* Such models may be more or less sophisticated, depending on the number of variables involved. Typically, they can be developed on a spreadsheet by simply aging the current workforce under currently prevailing promotion, separation, retirement, accession (including lateral), rotation, and career development practices. Box 3.3 describes such a model, for the acquisition workforces in the military services and defense agencies.

While an appropriate workforce projection model could be developed centrally for AFMC as a whole, we recommend that each center and laboratory directorate develop its own model because both the projection parameters and the competency variables of interest may differ among them. Moreover, as noted in the introduction, laboratory and center commanders now "own" their staffs and are responsible for management of their workforces. This modular approach, however, still needs to be aggregated at the HQ AFMC level to develop the supply profile needed for comparison against future workforce requirements (see Chapter Four).

Box 3.3
Workforce Projection Model

RAND developed a simple workforce aging model using a Microsoft Excel workbook to estimate how the longevity-related characteristics of a workforce are likely to change over a period of several years. The model also can be used to estimate how alternative human resources programs or policies affect hiring, retention, and/or promotion rates.

The model starts with a beginning inventory of the workforce, distributed by year of service from one to as many years as desired. The model uses continuation rates to calculate the number of workers in each year of service who are expected to remain in the workforce for an additional year. Workers who do not continue from one year to the next are counted as losses. The model then determines the number of gains (new hires) necessary to replace losses while accounting for any change in the target end strength. The model can also test how retention, accession, lateral hiring, and other potential personnel policy changes may affect these rates. Because the parameters differ between military and civilian personnel, separate models are needed.

As the model is currently aggregated, that is, it focuses on the force as a whole without distinguishing among occupations or other worker characteristics of interest. However, this workbook modeling approach can be readily expanded to accommodate as fine a breakdown of the workforce as desired.

For more details on this model, see Emmerichs, Marcum, and Robbert (2004).

CHAPTER FOUR

Comparing the Demand with the Supply

The outcomes of the previous steps in the process provide profiles of both the current and future competencies AFMC positions *need* (the demand) and the current and projected competencies the workforce *has acquired* (the supply). The next step uses this information to (1) compare the demand with the supply to identify potential gaps between the current and desired workforces and address other workforce issues and (2) assess alternative strategies for addressing gaps and ongoing workforce issues and support force management decisionmaking (Figure 4.1).

Management has three primary levers for filling gaps:

1. new accessions into the workforce (entry or lateral transfer)
2. changing the mix of military, civilian, and contractor personnel
3. force development through successive job assignments and education and training.

In addition to HQ AFMC, the centers and laboratory directorates should also analyze the characteristics of their respective workforces.

Figure 4.1
Compare Demand with Supply

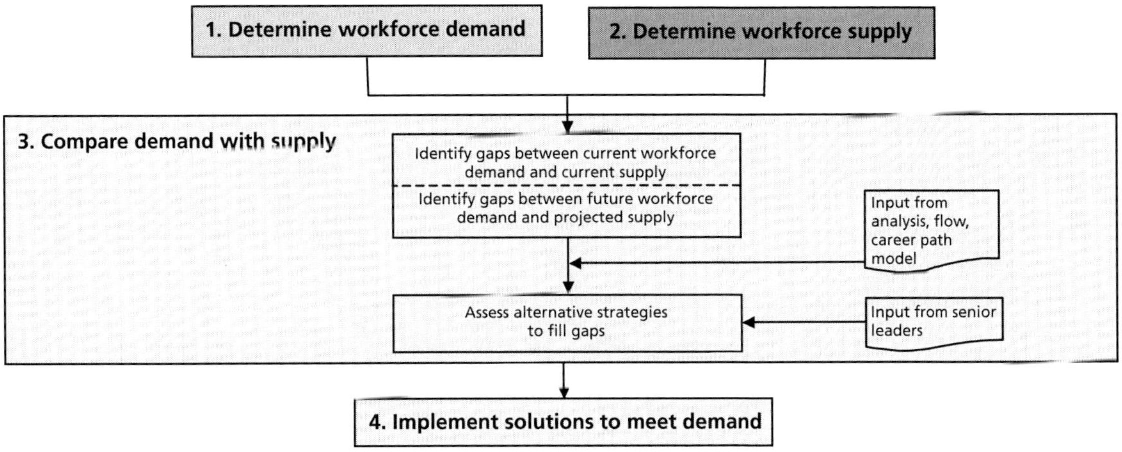

RAND *TR408-4.1*

Identifying Gaps and Other Workforce Issues

Several analytical activities need to be performed for effective force management:

- identify gaps between job demand and supply
- monitor ongoing changes in the workforce
- address specific questions or issues raised by leadership
- assess career development activities and outcomes.

We illustrate below the kind of questions that the information emerging from the previous steps allows. These questions can be asked at the AFMC-wide level or at each of the respective centers and laboratory directorates, as well as at the level of smaller units, such as groups and even squadrons.

Identifying Gaps Between Supply and Demand

This analysis should be designed to identify discrepancies between the competencies positions require and the competencies the workforce has, for both current and projected competencies. At a minimum, it should address the following types of questions:

- Are there gaps between demand and supply for any specific competencies at any grade level AFMC-wide or within any center or laboratory directorate?
- Are there gaps between demand and supply in the combination of competencies required at any grade level AFMC-wide or within any center or laboratory directorate?
- Are there gaps between the competencies jobs need and the competencies the officers or civilians that are assigned to these jobs have acquired?

Table 4.1 illustrates the outcomes of such analyses in the case of officers in the space and missile (AFSC 13S) career field.

Monitoring Ongoing Changes in the Workforce

The workforce change can be expected to be ongoing as different organizations and individuals make different, often uncoordinated, hiring, promotion, separation, and assignment decisions. The cumulative results of these individual decisions may or may not be consistent with AFMC's and its constituent organizations' needs. Relevant questions to ask from the data include the following:

1. Are hires made during the last six months or year consistent with filling gaps in workforce competencies?
2. Are there any significant changes in pattern of accession, hiring, promotion, and assignment? Are these consistent with current and/or estimated future workforce requirements?
3. Are occupational and skill mix changes consistent with current and future workforce requirements?

Table 4.1
Gaps Between Supply and Demand for Officer Workforce in the Space and Missile (AFSC 13S) Career Field (%)

	Grade		
Background	**Major**	**Lt Col**	**Col**
AFS prefix			
R = War planner	–4	–2	–14
S = Safety	–2	–1	–5
W = Weapons and tactics	–8	–4	–6
Functional			
Requirements	—		–15
Research and development	—	–2	–11
Acquisition	—	–2	–5
Test and evaluation		–2	–13
Organization			
14th Air Force	–4	–7	14
20th Air Force	–5	—	–2
AFSPC	—	–7	–26
National Reconnaissance Office	–3	–10	–10
Space and Missile Center	—	—	–4
Space Warfare Center	—	–4	–6
Air Staff	–9	—	–7
Office of Joint Chiefs of Staff or Office of Secretary of Defense	–7	–1	—

SOURCE: Vernez et al. (2006).

NOTE: Only competencies with a deficit between the 13S officer supply and the 13S job demand are listed in the table. A dash indicates that the supply equals or exceeds the demand. The table indicates, for instance, that 4 percent fewer O-4 officers acquired contingency or war planner experience than is required by 13S positions. Similarly, 26 percent fewer O-6 officers had acquired AFSPC experience than is required by 13S positions.

Addressing Questions Raised by Leadership

Workforce management is a daily activity, with commanders, line managers, and functional managers often needing basic information on issues that they identify or that may have been brought to their attention. Senior leaders may also need to get a status report on specific activities or on progress made addressing issues they may be concerned with. Examples of such questions include

- Who received what type of training in the past year?
- How many staff members are being rotated across laboratory directorates, centers, or test units on an annual basis? What are the grades and competencies of these individuals?
- How many staff members are likely to retire within a specified period?

- Who has the necessary competencies to address a pressing issue or an unusual increase in workload?
- Which occupations are understrength, and which may be overstrength and in which organizations?

Figure 4.2 illustrates the outcome of an analysis that addressed the last question above for officers in the acquisition management career field. This figure indicates that about three times more lieutenants were assigned to this career field than it required and that it was consistently undermanned at higher grades, leaving a gap in the number of those with the experience needed to do the job. However, it also shows that the requirements could be met at the O-3 to O-6 levels if all acquisition officers were assigned to acquisition positions instead of to student-transient-patient-prisoner, special duty, or outside positions.

Assessing Career Development

With the new Air Force and AFMC emphasis on force development, a new set of questions will need to be addressed to monitor the outcome of the process and assess its effectiveness. Questions of interest are likely to include the following:

Figure 4.2
Understrength Career Field: Acquisition Management

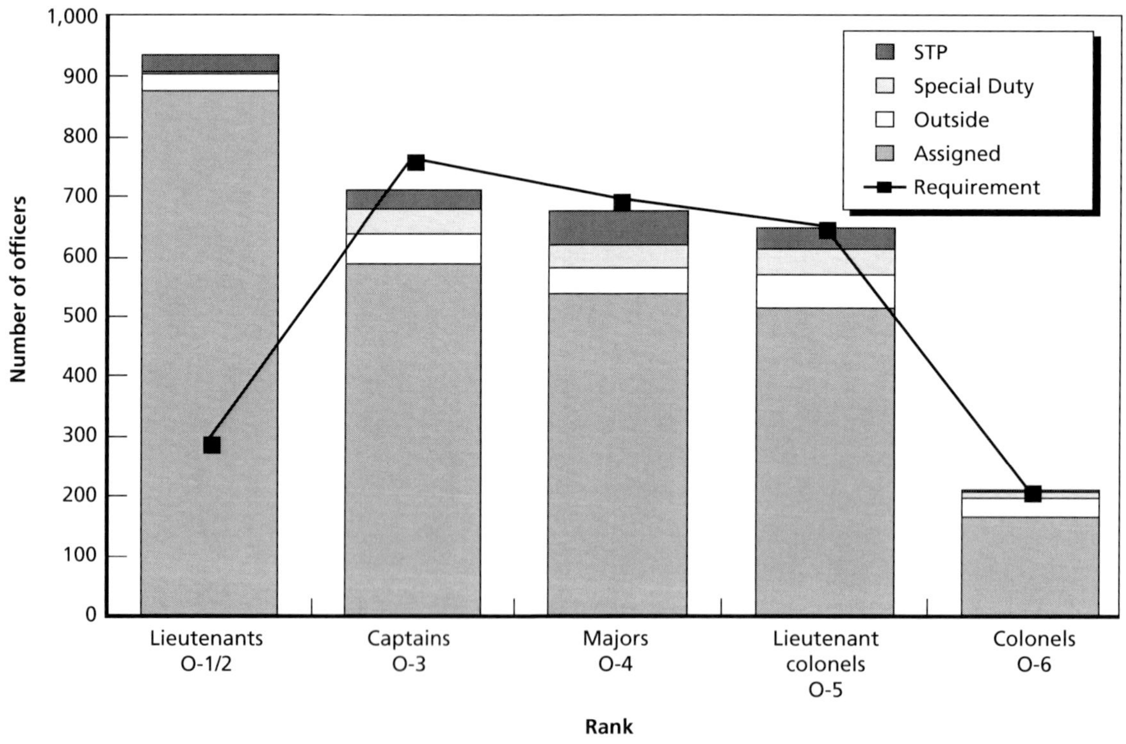

RAND *TR408-4.2*

- What guidelines were given to the development teams? Are they consistent with AFMC workforce requirements?
- Who and how many participate in leadership development programs? What are the competencies of participants?
- Are individuals appropriately rotated across jobs and organizations? What are the competencies of those being rotated? Are rotations consistent with workforce requirements?

Policy Decisions Needed. None are needed.

Methods, Data, and Tools. The data files developed in steps 1 and 2 (Figure 4.1) provide the basis for addressing the types of questions listed above. The analytical tools needed range from simple comparison of percentage distributions to cluster analysis to modeling—the tools statisticians, operations researchers, econometricians, quantitative sociologists, and behavioral scientists typically use.

Assignment of Responsibilities. Ideally, both HQ AFMC and each center and laboratory directorate should have some capacity to conduct ongoing analyses of their respective workforces because the relevant force management questions are likely to differ across them. As noted in the summary, however, this capacity is not currently available in many or all of these business units and will need to be developed if AFMC and its centers and laboratory directorates are to engage in effective workforce planning and development.

Assessing Alternative Strategies to Fill Gaps

Three questions needs to be addressed in this task:

1. Is it feasible to develop a workforce that fully meets the current or the future job competencies needed under prevailing hiring, promotion, separation, rotation, and career development policies?
2. What changes in hiring, promotion, separation, training, rotation, and career development policies may be needed to alleviate current gaps in the short and long terms?
3. How many staff are needed at each grade level to meet the leadership requirements at the top, and what combination of competencies do they need to accumulate?

Policy Decisions Needed. Two types of leadership decisions are needed. First, analysts need guidance on the range of policy options they may consider to fill the gaps identified. The gaps could be filled by anyone or by a combination of hiring, separation, education and training, and force development strategies or programs. Second, once the leadership has received appropriate analytical information about feasibility and costs of alternative options for closing the gap, it needs to make a decision on which to actually implement.

Methods, Data, and Tools. Addressing these questions requires developing comprehensive or partial models that simulate the flow of officers and civilians through grades and assignments throughout their careers, keeping track of the competencies the jobs require and the competencies staff members acquire in these jobs. Such models include policy variables—typically, hiring, promotion, rotation, and separation rates—that may be disaggregated by type of competency to assess the effects of alternative policies. The architecture of such a model, which RAND used in its study of the 13S career field, is shown in Figure 4.3.

The development and use of such models requires specialized training and is time consuming—time that may not always be available. While modeling is extremely useful, simpler forms of analysis and deductive analysis may also have to be used to assess different options.

Assignment Responsibilities. These would be as for the previous task.

Figure 4.3
Structure of Flow Model for Analyzing Force Development Options for the Space and Missile (AFSC 13S) Officer Career Field

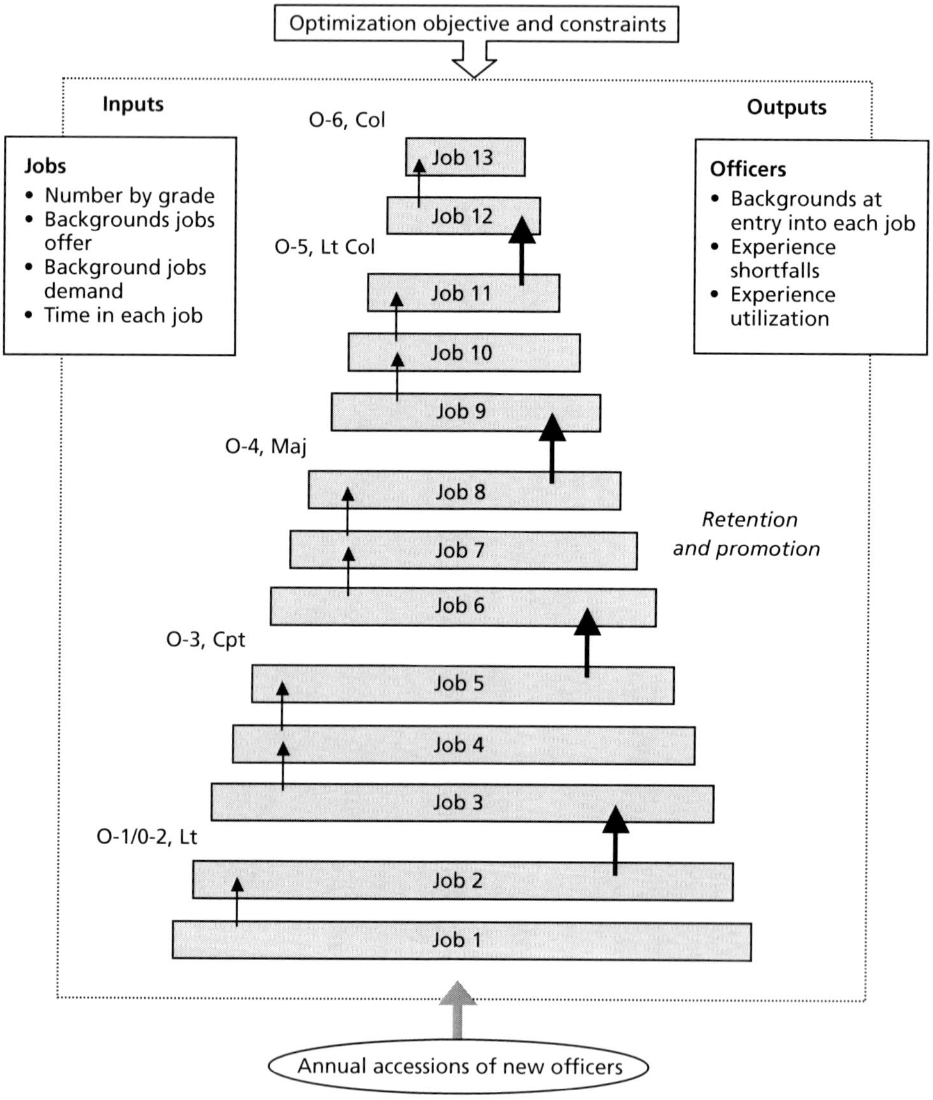

RAND *TR408-4.3*

CHAPTER FIVE

Implementing Solutions to Meet Workforce Requirements

The analytical steps described in the previous chapters provide the information needed to make periodic decisions about the number of military, civilian, and contractor staff that ought to be hired, separated, promoted, rotated, and/or trained and developed in a specified way. These decisions, in turn, need to be translated into a set of guidelines for the organizations and staff charged with implementing them.

Some of the guidelines that need to be provided to implement the force requirements defined in earlier steps are relatively straightforward (Figure 5.1). Such is the case for hiring requirements, via either accession or lateral hires, for civilians and contractor personnel. However, creating force development guidelines for both military and civilian personnel is more complex because decisions on training, promotion, and job assignments must be made for each individual separately; at the same time, the cumulative sum of these individual decisions must eventually aggregate to the requirements for the AFMC workforce as a whole.

Guidelines for Recruitment and Hiring Decisions

To guide accessions, annual or biannual targets will need to be set for the number of individuals and with what competencies that should be recruited or hired at the entry level (and for civilian and contractor personnel at the more senior levels) for each organization.[1] These targets must then be conveyed to the staff responsible for recruiting and hiring.[2] Absent such guidelines, supervisors may tend to hire replacements that preserve existing distributions of competencies. Better guidelines can orient hiring toward future rather than past needs.

AFMC's ability to meet its targets may be limited by the capacity of the hiring entities to identify qualified persons with the required competencies. Hence, it is important that AFMC and each organization within it regularly monitor actual hiring against the targets. If new hires appear to diverge from the targets, remedial actions may be necessary.

---

[1] Although in most cases, setting biannual targets is adequate, the period for which hiring targets are set may be shorter to meet operational needs.

[2] Military personnel recruitment is centralized, and AFMC currently has limited leverage in assuring that the number and types of new recruits meet its occupational requirements now and for the future. Galway et al. (2005) argues that the Air Force needs to make personnel decisionmaking process changes to balance its force requirements across career fields.

Figure 5.1
Implement Solutions to Meet Demand

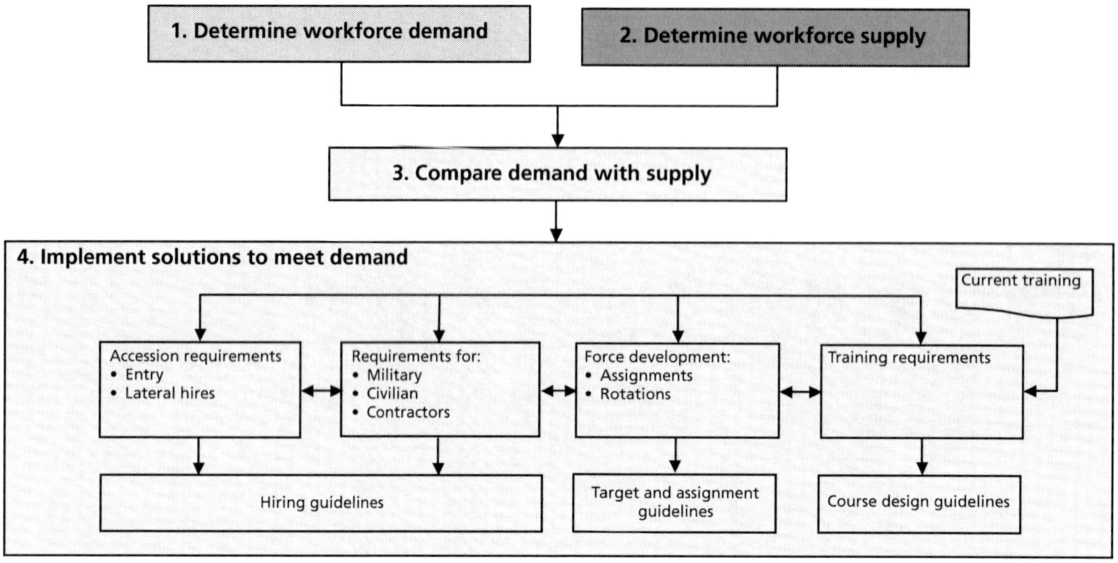

RAND *TR408-5.1*

Guidelines for Force Development

As part of its force development program, the Air Force has institutionalized career development teams (DTs) for both military and civilians. The teams' mission is to make decisions on education, training, and job assignment that account for individual preferences and that are consistent with Air Force–wide force requirements. Because AFMC is represented (or can be represented) on the corporate DTs, it makes sense for AFMC to use them to ensure that its force requirements are met along with those of the Air Force.

To the extent that corporate DTs do not cover all personnel relevant to AFMC force management needs, which is likely, AFMC should establish AFMC-specific DTs whose mission would be similar to that of the corporate DTs but that would focus on personnel not covered by the corporate effort. Whether these AFMC-specific DTs ought to be at the business unit level or AFMC-wide will depend on the decisions AFMC makes on competencies of interest only to AFMC and on the need for rotation of assignments across its organizations. Our expectation would be for broader, AFMC-wide DTs.

Whether Air Force–wide or AFMC-specific, the DTs themselves need guidelines both on the number of individuals (military and civilians) that need to be developed over time with a specified set of competencies and on targets for the number of individuals that ought to have specified competencies and types of training at each grade. As yet, there is no experience with developing such guidelines. The form and the specificity with which these guidelines may have to be developed will in large measure depend on the outcomes of the previous three steps.

Managing Ongoing Education and Training Activities

Currently, a broad range of education and training activities is available to or required of AFMC workforces. These offerings may be sponsored corporately or functionally at Air Force, AFMC, center, or local levels. The AFMC commander asked AFMC/A1 to consider approaches for rationalizing the management of the many offerings, and AFMC/A1, in turn, asked PAF to provide recommendations regarding its possible roles.

In our view, a first step would be to survey the extent of the current education and training offerings available to AFMC workforces to gather information on their sponsors, content, throughput capacity, participants, selection criteria, delivery mode, and costs. Beyond that, AFMC/A1's roles in managing training and education programs should vary as a function of the level and type (corporate or functional) of sponsorship of the offering. Table 5.1 contains our recommendations.

Table 5.1
AFMC/A1 Roles in Managing Education and Training Programs

Level of Sponsorship	Type of Sponsorship	Recommended Roles
Air Force	Corporate	As appropriate per program policies. May include determining throughput requirements for AFMC workforce participation (through planning processes advocated in this document), negotiating for quotas to meet AFMC needs, nominating or selecting individuals for attendance, and advocating funding levels to support AFMC workforce participation if not centrally funded.
	Functional	Very limited; AFMC functional managers provide the primary interface with Air Force functional counterparts.
AFMC	Corporate	Identifying unmet needs (through planning processes advocated in this document). Evaluating essentiality. Determining throughput requirements and curriculum content, selecting participants or establishing guidelines for attendance, and advocating funding levels.
	Functional	Evaluating essentiality. Advising functional managers on identifying needs, throughput requirements, curriculum content, and selection criteria. Advocating funding.
Center or local	Corporate	Evaluating essentiality. Advising line managers on identifying needs, throughput requirements, curriculum content, and selection criteria. Advocating funding.
	Functional	Evaluating essentiality. Advising functional managers on identifying needs, throughput requirements, curriculum content, and selection criteria. Advocating funding.

CHAPTER SIX

Conclusions and Recommendations

As a contribution to AFMC's effort to establish and maintain a capability to engage in proactive, analytically based workforce planning and development, this report has described the policy decisions and types of analyses involved and has recommended an allocation of responsibilities among AFMC organizational components. Before it can engage in strategic workforce planning and force development, however, AFMC needs first to acquire an analytical capability both at HQ AFMC and in its centers and laboratory directorates, a capability that is not present today. This capability can be built incrementally over time.

Consistent with AFMC's concept of operations, we generally recommend that the responsibilities for workforce planning and development be assigned to the business units—its product and logistics centers and its laboratory directorates. The primary roles of HQ AFMC should be to guide the process and provide support and assistance to the business units and to moderate the workload and resource trade-offs that may need to be made among its business units.

AFMC does not need to focus its planning and force development effort on its entire force. It should identify the organizational elements that are critical to its mission and for which it will strategically manage the workforce. As noted above, we recommend a focus on AFMC core business units, the centers and laboratories that account for about 70 percent of its workforce. Force management for other activities, including base support, can be done effectively within the context of Air Force–wide processes and programs. Within its core businesses, AFMC also needs to identify the positions that are critical to their respective missions. Such positions should be identified in all grades or levels and for military, civilian, and contractor personnel.

Another key decision that needs to be made up front concerns the set of competencies that AFMC ought to concentrate on acquiring and/or developing. Typically, desired competencies will include not only technical skills but also product systems, combat, intra- and interservice organizational experience, and leadership and management skills.

Making these decisions and then applying the workforce planning and development process outlined in these pages will require the active and frequent involvement of AFMC leadership and of its center and laboratory commanders.

APPENDIX A

Sample Worksheets

Figures A.1 and A.2 illustrate worksheets that can be used to gather information on required competencies. Figure A.3 illustrates a spreadsheet to use in estimating future required competencies and rating the element of change.

Figure A.1
Worksheet Used to Gather Competency Required for Space and Missile (AFSC 13S) Positions

Position group: ______

Columns: Critical, Important, Useful (check); Length, Recency (years)

1. Operational experience

- a A = Satellite C2
- b B = Spacelift
- c C = Missile combat crew
- d D1 = Space surveillance, ground
- e D2 = Space surveillance, orbital
- f E1 = Space warning, ground
- g E2 = Space warning, orbital
- h Other: ______
- i Other: ______

2. Previously-held AFS prefix

- a C = Commander
- b B = Squadron Ops Officer
- c K = Instructor
- d Q = Stan/Eval
- e R = Contingency/War Planner
- f S = Safety
- g T = Formal Training Instructor
- h V = Automated Fctnl Appl Anal
- i W = Weapons & Tactics Instr
- j V = Automated Fctnl Appl Anal
- k X = Nonrated Aircrew
- l Y = Analytic Studies Officer
- m Other: ______
- n Other: ______

3. Prior functional experience

- a Personnel (e.g., DP, A-1, J-1)
- b Intelligence (e.g., IN, A-2, J-2)
- c Current ops (e.g., XO, DO, A-3, J-3)
- d Logistics (e.g., LG, IL, A-4, J-4)
- e Plans, programs (e.g., XP, A-5, J-5)
- f Communications (e.g., SC, A-6, J-6)
- g Requirements (e.g., XR, DR)
- h R&D
- i Acquisition
- j Test & evaluation
- k Contracting
- l Financial management
- m Pol-Mil
- n Education/Training
- o Mgt of civilians, contractors
- p Other: ______
- q Other: ______

4. USAF school (unique to position)

- a Technical: ______
- b PME: ______
- c ______

5. Degree area (unique to position)

- a Undergrad: ______
- b Graduate: ______

6. Command experience

- a Flight
- b Squadron
- c Ops group
- d Support group
- e Logistics group
- f Any group
- g Wing
- h NAF
- i Joint
- j Center
- k School
- l Other: ______
- m Other: ______

7. Organizational experience

- a Group-level
- b Wing-level
- c 14th AF
- d 20th AF
- e Other NAF
- f CAF (ACC, USAFE, PACAF)
- g AFSPACE
- h AETC
- i Air Staff
- j AIA
- k NRO
- l JCS
- m DTRA
- n SMC
- o USSTRATCOM
- p USSPACECOM
- q Other: ______
- r Other: ______

8. DAL occupational certification

- a Space core, Level 1 (basic)
- b Space core, Level 2 (senior)
- c Space core, Level 3 (master)
- d Space application, Lvl 1 (basic)
- e Space application, Lvl 2 (senior)
- f Space application, Lvl 3 (master)

9. Combinations or other backgrounds

- a ______
- b ______
- c ______

Can this position be filled by an officer of a grade lower than the one authorized for the position? YES NO

RAND *TR408-A.1*

Figure A.2
Worksheet Used to Gather Competency Required for Air Force Rated Officer Positions

Prior Experience, Education, and Training That May Be Needed for One or More USAF Rated Positions Above Company Grade

1. Operational experience
- a 1-character RDTM
- b 2-character RDTM
- c RDTM(s)________________
- d 2-character AFSC
- e 3-character AFSC
- f 5-character AFSC
- g AFSC(s)________________
- h Training Instructor pilot/nav
- i EWO
- j WSO
- k ALO
- l TALO
- m UAV
- n SOF
- o 5yr currency
- p 7yr currency
- q Experience________________
- r Other: ________________

2. Previously-held AFS prefix
- a A = Op War Instructor
- b B = Squadron Ops Officer
- c C = Commander
- f F = Flight Eval
- j J = Parachutist
- k K = Instructor________________
- l L = Life Support
- q Q = Stan/Eval
- r R = Contingency/War Planner
- s S = Safety
- t T = Formal Training Instructor
- u U = Info Ops
- v V = Automated Fctnl Appl Anal
- w W = Weapons & Tactics Instr
- x X = Nonrated Aircrew
- y Y = Analytic Studies Officer
- z Other: ________________

3. Prior functional experience
- a Any functional experience
- b Current ops (e.g., XO, DO, A-3, J-3)
- c APE*: Combat Ops
- d APE*: Mobility Ops
- e APE*: Info/Recce Ops
- f Info warfare
- g Logistics (e.g., LG, IL, A-4, J-4)
- h Maintenance
- i Inspection
- j Safety
- k War planner
- l Intelligence (e.g., IN, A-2, J-2)
- m Space
- n Test & evaluation
- o Education
- p Plans, programs (e.g., XP, A-5, J-5)
- q Requirements (e.g., XR, DR)
- r R&D
- s Acquisition
- t Financial management
- u Mgt of civilians, contractors
- v Contracting
- w Communications (e.g., SC, A-6, J-6)
- x Pol-Mil
- y Personnel (e.g., DP, AFPC, A-1, J-1)
- z Medical
- aa Training
- bb Other: ________________

*Note: APE = Aerospace Power Employment

6. Combinations/Other backgrounds
- a ________________
- b ________________
- c ________________
- d ________________
- e ________________
- f ________________

7. Grade
- a Must hold the authorized grade

4. Organizational experience
- a Any staff
- b Wing/Group staff or OSS
- c AOC/CAOC/JAOC
- d NAF Staff or above
- e MAJCOM Staff or above
- f Any staff above MAJCOM
- g NAF
- h Any MAJCOM Staff
- i MAJCOM Staff appropriate to mission
  - j ACC
  - k AMC
  - l AFMC
  - m AFSPC
  - n AETC
  - o ANG/AFRC
  - p USAFE
  - q PACAF
  - r AFSOC
- s Any DRU or FOA
  - t USAFA
  - u AFPC
  - v AF Test and Evaluation Center
  - w AIA
- x Warfare center
- y Tanker Airlift Control Center
- z Systems Center
- aa Other center________________
- bb Air Staff
- cc Defense agencies
- dd Any joint staff
  - ee OSD staff
  - ff JCS
  - gg Any geog command staff
  - hh Any combatant command staff
  - ii JFCOM
  - jj TRANSCOM
  - kk STRATCOM
  - ll SOCOM
  - mm EUCOM
  - nn PACOM
  - oo CENTCOM
  - pp SOUTHCOM
  - qq NORTHCOM
- rr Other: ________________

5. Prior leadership experience
- a Supervisory
- b Flight command
- c Any command
- d Squadron/CC
  - e Operational squadron
  - f Training squadron
  - g Test squadron
- h Group/CC
  - i Ops group
  - j Logistics group
  - k Support group
- l Wing/CC
  - m CAF wing
  - n MAF wing
  - o C2ISR wing
  - p SOF wing
  - q Training Wing
  - r Air Base/Other Wing
- s Expeditionary Command
- t NAF
- u Joint
- v Center
- w School/Training
- Other: ________________

8. Degree area (unique to position)
- Undergrad: ________________
- Graduate: ________________

9. PME/Tech training (unique to position)
- a Technical: ________________
- b Technical: ________________

PME (highest level):
- c ASBC ________________
- d SOS ________________
- e IDE ________________
- f SDE ________________
- g Other________________
- h In-Residence________________

RAND *TR408-A.2*

Figure A.3
Sample Spreadsheet for Estimating Future Job Requirements by Element of Change

Civilian Functional Requirements by Key Changes (Notional)

	FY05	OFFICER-LEVEL														Total Officers	ENLISTED-LEVEL					Total Enlisted
		ACQUISITION		CONTRACTING	FINANCE		LOGISTICS	MAINTENANCE	ENGINEERING						OTHER		CONTRACTING	FINANCE	INFO MGT	LOGISTICS	OTHER	
Current Authorizations	Specialty code & description	060cx program director	063ax acquisition manager	064px contracting	065fx financial mgmnt	065wx cost analysis	021rx logistics readiness	021ax acft maintenance	033sxa comm and info, eng	062exa devlmt engr, aero	062exc devlmt engr, comp sys	062exe devlmt engr, electr	062exg devlmt engr, project	062exh devlmt engr, mechan	Otheer officer specialties		6c0x1 contracting	6f0x1 fin mgmt/comp	3a0x1 info management	2g0x1 logistics plans	9s1 tech appl specl	
	CIV	1	74	60	23	13	18	0.3		8	1	41	22	3	8	265	2	0	10	0	1	13
	%	0%	28%	23%	9%	1%	7%	1%	0%	3%	0%	15%	8%	1%	3%	100%	15%	0%	77%	0%	8%	100%
Unit	ASC - Reconnaissance Wing (notional)																					

For each key change, estimate effect on civilian distribution functions.

Product change 1																					
# by specialty==>																					
% by specialty==>																					
Product change 2																					
# by specialty==>																					
% by specialty==>																					
Process change 1																					
# by specialty==>																					
% by specialty==>																					
Process change 2																					
# by specialty==>																					
% by specialty==>																					
Product change 1																					
# by specialty==>																					
% by specialty==>																					
Organiz. change 2																					
# by specialty==>																					
% by specialty==>																					

RAND *TR408-A.3*

APPENDIX B

Protocol Used to Identify Elements of Change for the AFMC Armament Enterprise

1. What major munitions changes are taking place or do you anticipate over the next 10 to 15 years or so (e.g., the integration of directed energy into weaponry, the need for greater precision in the delivery of weapons)?

   For each munitions change that is identified:

   - What are the principal differences between the new or enhanced munitions program and the Armament Enterprise current munitions programs?
   - Will the new program require changes in the Armament Enterprise acquisition business process? If so, what kinds of changes?
   - Will the new program require a different mix of skills at the Armament Enterprise? If so, how will the current mix of skills need to change to support the new program?

2. What functions have been added or do you anticipate will be added or eliminated over the next 10 years or so?

   For each function added or eliminated that is identified:

   - How many jobs will be affected?
   - What is the mix of skills needed or affected?

3. What major organizational changes are you making or do you anticipate in the Armament Enterprise over the next 10 or so years (e.g., PEO restructuring, establishment of the Development Test and Evaluation Center, movement of the AAC Enterprise organization from under the Deputy for Acquisition to under the AAC/CC, reconfiguration of SPOs as Capability SPOs)?

   For each change that is identified:

   - How will the organizational change influence the way the acquisition business is performed at the AAC?
   - Will the organizational change affect the mix of skills needed to accomplish the future objectives of the AAC?

4. Do you anticipate major changes in the acquisition process over the next 10 years or so (e.g., due to need to develop integrated munitions support or systems of systems products or due to joint acquisition requirements)?

For each change that is identified:

- How will the change affect the way the acquisition business is performed in the Armament Enterprise?
- Will the organizational change affect the mix of skills needed to accomplish the future objectives of the Armament Enterprise?

5. What changes in the AAC customer base do you anticipate occurring in the next 10 years?
   - What is the underlying reason for this change?
   - Will the acquisition business process of the AAC need to change to manage programs required by the new customer(s)? If so, how will it need to change?
   - Will a different mix of skills be needed to meet the needs of the new AAC customer(s)?
6. Do you anticipate major changes in the extent to which the AAC relies on prime and support contractors?
   - How would the change affect the acquisition business process of the AAC?
   - Will the change affect the mix of skills required to do business at the AAC?
7. The acquisition workforce is expected to decrease by 8 percent by 2008. How do you anticipate this reduction will be taken? Which organizations and/or functions will be affected?
8. Are there other changes that you can foresee that are likely to have an effect on how the AAC performs its acquisition business or on the mix of skills needed for the Armament Enterprise?

References

Air Force Materiel Command, *Roles and Responsibilities of AFMC Center Line Direct Reports and Functionals*, November 7, 2004.

Asch, Beth, and James R. Hosek, *Looking to the Future: What Does Transformation Mean for Military Manpower and Personnel Policy?* Santa Monica, Calif.: RAND Corporation, OP-108-OSD, 2004. Online at http://www.rand.org/pubs/occasional_papers/OP108/.

Crawford, Jessica, "State Workforce Planning 2000," Syracuse, N.Y.: The Maxwell School's Alan K. Campbell Public Affairs Institute, Government Performance Project, 2001.

Dewar, James A., *Assumptions-Based Planning: A Tool for Reducing Avoidable Surprises*, New York: Cambridge University Press, 2002.

Dewar, James A., Carl H. Builder, William M. Hix, and Morlie H. Levin, *Assumption-Based Planning: A Planning Tool for Very Uncertain Times,* Santa Monica, Calif.: RAND Corporation, MR-114-A, 1993. Online at http://www.rand.org/pubs/monograph_reports/MR114/.

Emmerichs, Robert M., Cheryl Y. Marcum, and Albert A. Robbert, *An Operational Process for Workforce Planning*, Santa Monica, Calif.: RAND Corporation, MR-1684/1-OSD, 2004a. Online at http://www.rand.org/pubs/monograph_reports/MR1684.1/.

Emmerichs, Robert M., Cheryl Y. Marcum, and Albert A. Robbert, *An Executive Perspective on Workforce Planning*, Santa Monica, Calif.: RAND Corporation, MR-1684/2-OSD, 2004b. Online at http://www.rand.org/pubs/monograph_reports/MR1684.2/.

Galway, Lionel A., Richard T. Buddin, Michael R. Thirtle, Peter S. H. Ellis, and Judith D. Mele, *Understrength Air Force Officer Career Fields: A Force Management Approach*, Santa Monica, Calif.: RAND Corporation, MG-131-AF, 2005. Online at http://www.rand.org/pubs/monographs/MG131/.

Gates, Susan, Christine Ebner, and Edward G. Keating, *Civilian Workforce Planning in the Department of Defense,* Santa Monica, Calif.: RAND Corporation, MG-449-OSD, 2006. Online at http://www.rand.org/pubs/monographs/MG449/.

Gates, Susan M., and Albert A. Robbert, *Personnel Savings in Competitively Sourced DOD Activities*, Santa Monica, Calif.: RAND Corporation, MR-117-OSD, 2000. Online at http://www.rand.org/pubs/monograph_reports/MR1117/.

Robbert, Albert A., Stephen M. Drezner, John Boon, Larry Hauser, Craig Moore, Lynn Scott, Herbert Shukiar, *Integrated Planning for Air Force Senior Leader Workforce: Background and Methods*, Santa Monica, Calif.: RAND Corporation, TR-175-AF, 2005. Online at http://www.rand.org/pubs/technical_reports/TR175/.

Sullivan, John, "Why You Need Workforce Planning," *Workforce*, Vol. 81, No. 12, 2002, pp. 46–50.

Vernez, Georges, Craig Moore, Steven Martino, and Jeffrey Yuan, *Improving the Development and Utilization of Air Force Space and Missile Officers*, Santa Monica, Calif.: RAND Corporation, MG-382-AF, 2006. Online at http://www.rand.org/pubs/monographs/MG382/.

U.S. Air Force, *USAF Scientist and Engineer Future Study,* Washington, D.C., 2002.

U.S. Secretary of the Air Force, *Total Force Development Active Duty Officer*, Air Force Instruction 36-2640, Vol. 1, January 23, 2004. Online at https://www.dp.hq.af.mil/dpx/dpxf/docs/afi36-2640v1.pdf (as of August 22, 2006).

U.S. Office of Management and Budget, *Revised Supplemental Handbook: Performance of Commercial Activities*, Washington D.C., Circular A-76, 1996. 2003 revision available online at http://www.whitehouse.gov/omb/circulars/a076/a76-rev2003.pdf.

Winkler, John D., Herbert J. Shukiar, James A. Dewar, Matthew W. Lewis, Beth A. Benjamin, Jerry M. Sollinger, John E. Peters, and Harry Thie, *Future Leader Development of Army Noncommissioned Officers*, Santa Monica, Calif.: RAND Corpoation, CF-138-A, 1998.